Vehicular Networks - Principles, Enabling Technologies and Modern Applications

Edited by Abdelfatteh Haidine

Published in London, United Kingdom

Vehicular Networks - Principles, Enabling Technologies and Modern Applications
http://dx.doi.org/10.5772/intechopen.102303
Edited by Abdelfatteh Haidine

Contributors
Abdelfatteh Haidine, Abdelhak Aqqal, Abdelmoula Ait-Allal, Ayoub El Idrissi, Aziz Dahbi, Binod Kumar Pattanayak, Chen Sun, George Kamata, Hironao Kawashima, Hisafumi Kokubugata, Lantao Li, Mariyam Ouaissa, Mariya Ouaissa, Omkar Pattnaik, Ryotaro Fukui, Sasmita Pani, Tao Cui, Zakaria Boulouard, Zhaoyu Zhang

First published in London, United Kingdom, 2024 by IntechOpen
IntechOpen is the global imprint of INTECHOPEN LIMITED, registered in England and Wales, registration number: 11086078, 167-169 Great Portland Street, London, W1W 5PF, United Kingdom

British Library Cataloguing-in-Publication Data
A catalogue record for this book is available from the British Library

Additional hard and PDF copies can be obtained from orders@intechopen.com

Vehicular Networks - Principles, Enabling Technologies and Modern Applications
Edited by Abdelfatteh Haidine
p. cm.
Print ISBN 978-1-80356-872-0
Online ISBN 978-1-80356-873-7
eBook (PDF) ISBN 978-1-80356-874-4

Meet the editor

Dr. Ing. Abdelfatteh Haidine is Professor Habilité at the University of Applied Sciences, El Jadida, Morocco. He is a member of the Laboratory of Information Technologies at the same university. His research interests include the simulation and evaluation of wireless and mobile networks, as well as design (network planning/migration and resources allocation) and evaluation of communication networks for smart environments, especially smart cities, smart grids, smart maritime ports, and smart environment monitoring. He holds a Ph.D. from Technische Universität Dresden, Germany. His Ph.D. studies focused on the planning and optimization of communication networks utilizing combinatorial optimization and multi-objective optimization/ multi-criteria decision-making. Dr. Haidine has worked on different European research and industry projects as well as with various companies as a consultant for intelligent network communications, with a focus on the planning of communication solutions for power utilities to build smart grid communication platforms. His tasks covered technical specifications, planning and performance evaluation of communication solutions, as well as techno-economics aspects for smart grid projects.

Contents

Preface

Vehicles play a crucial role in modern society, contributing significantly to various aspects of life. Vehicles provide mobility for people and goods, facilitating travel and transportation of goods over short and long distances. They enable commuting to work, school, healthcare facilities, and recreational activities, thus enhancing overall accessibility and quality of life. The automotive industry is a major contributor to the global economy. It creates millions of jobs worldwide, directly and indirectly, in manufacturing, sales, services, and related sectors. Additionally, vehicles support other industries by transporting raw materials and finished goods, fostering economic growth and development.

Vehicles ushered in the era of autonomy, first through assisted driving and second with the goal of reaching full autonomy in driving, parking, and so on. Vehicles face numerous challenges, ranging from technological limitations to social and environmental concerns. Addressing these challenges requires a holistic approach involving collaboration between governments, industry players, academia, and civil society to foster innovation, sustainability, and resilience in the automotive sector.

As part of this innovation, vehicular networks based on vehicular communications, also known as V2X (Vehicle-to-Everything) communication, refers to the exchange of data between vehicles (V2V), infrastructure (V2I), pedestrians (V2P), and other entities such as cyclists and road operators (V2R). This communication enables vehicles to interact with their surroundings, improving safety, efficiency, and convenience on the road.

This book discusses different aspects related to vehicular networks. It is organized into two sections. The first section, "Application Fields", begins with Chapter 1, "Introductory Chapter: Vehicles Everywhere, and Vehicular Communications Should Follow!". The introductory chapter gives an overview of the different and most recent and innovative applications of modern industries for vehicles, such as automated guided vehicles (AGV) in smart maritime ports or in logistics companies, and more. The main idea of this introduction is that where we have vehicles, we should also have vehicular networks, which are necessary to overcome some of the previously cited challenges. Chapter 2, "C-V2X Vision in the Chinese Roadmap: Standardization, Field Tests, and Industrialization", analyzes activities and applications of vehicular communications in the special case of China, which is a large country with a large population and huge number of used vehicles either in private lives or industries. China continuously and actively focuses on and contributes to C-V2X technology development. Starting from the standardization progress, the chapter presents several Chinese-specific use cases and communication messages for complicated urban traffic. Based on these higher-layer protocols, annual field tests are conducted to determine the interoperability among chipsets, modules, security certificates, and original equipment manufacturers (OEMs). Putting forth effort on industry standards and tests, China is fast commercializing C-V2X based road services (e.g., robotaxi, robot bus) in representative cities. Thus, in this chapter, we highlight advanced Chinese C-V2X philosophy. Chapter 3, "Mapping the Applications of Vehicular Communications in the Context of Smart Maritime Ports", discusses the very special use case of autonomous vehicles in smart maritime ports to load/unload or transport containers. In the smart maritime port environment, vehicular applications are adopted everywhere, such as AGVs to transport containers, unmanned aerial vehicles for different port operations, and more. This chapter discusses some concrete examples of these vehicular applications in the smart port environment and the adequate and optimal vehicular communication technologies to be deployed to support a reliable data transmission for these applications.

The second section, "Advanced Approaches", presents some advanced approaches to optimize the functioning of vehicular communications, such as machine learning as part of artificial intelligence. It also explores how to address some of the major challenges facing vehicular communications, like security issues. As vehicles become more connected and reliant on digital systems, they are increasingly vulnerable to cybersecurity threats such as hacking, malware, and data breaches. Safeguarding vehicle cybersecurity requires implementing robust security measures, including encryption, intrusion detection systems, and secure software development practices. Therefore, Chapter 4, "Secure Vehicular Networking: Architectures, Applications, Attacks, and Challenges", delves into the intricate details of vehicular communications, exploring their architectures, characteristics, applications, and governing standards. Furthermore, it elucidates the concept of communication security within these networks, outlining the existing models of attackers, classifications of attacks, and fundamental security mechanisms. The comprehensive overview sheds light on the critical aspects of securing vehicular networks in the ever-evolving landscape of communication technology. Machine learning (ML), as an integral part of artificial intelligence (AI), has recently been used in wireless networks to provide continuous information to deal with culturally challenging issues. Chapter 5, "VANET: A Machine Learning Approach", examines recent progress in the use of ML in automotive networks. It differentiates the obvious characteristics of multifunctional vehicle networks and promotes the use of ML to deal with subsequent difficulties. Chapter 6, "Speed Control of Inflow Vehicles for Merging Support on Motorways with Limited I2V Communication", discusses the speed control of inflow vehicles for merging support on motorways through the special scenario of automated driving in a limited V2I infrastructure environment.

Abdelfatteh Haidine
National School of Applied Sciences El Jadida (ENSAJ),
University Chouaib Doukkali,
El Jadida, Morocco

Section 1

Application Fields

Chapter 1

Introductory Chapter: Vehicles Everywhere, and Vehicular Communications Should Follow!

Abdelfatteh Haidine

1. Introduction—Vehicles everywhere!

The vehicles played a major role in the evolution of the human life and the industrial/modern society. Helping people to travel and carry merchandise over distances in a short time and in increasing comfort, they did not stop their evolution toward unbelievable and unprecedented manner. In the recent years, vehicles started assisting drivers in the driving, with the long-term objective to overtake the entire driving operations in an autonomous way to reach the full autonomous driving vehicles.

While autonomous driving vehicles are still competing to gain ground in the practice on public roads and fulfill the security requirements, autonomous vehicles have already been established in industrial applications. Different brilliant examples can be cited, as illustrated in **Figure 1**. Unmanned aerial vehicles (UAV) or drones are gaining increasing interest in different applications domains, such as the delivery tasks. The delivery items can be products purchased through online shopping or medical products in emergency situations, such as catastrophes/floods/etc., or accidents in difficult accessibly locations. A second example is the automated guided vehicles (AGV). These have a spread utilization in different industrial domains, such as the AGVs used to transport containers on maritime ports arrays or the AGVs used in the warehouses of giant online trade enterprises. Last but not least, the autonomous maritime vehicles, that is. vessels or ships, are new innovative solution to optimize the maritime transport of goods. The autonomy of the vessels can be one of four levels as defined by the International Maritime Organizations (IMO), as discussed in Ref. [1]:

- *Degree one: Ship with automated processes and decision support*: Seafarers are on board to operate and control shipboard systems and functions. Some operations may be automated and at times be unsupervised but with seafarers on board ready to take control.

- *Degree two: Remotely controlled ship with seafarers on board*: The ship is controlled and operated from another location. Seafarers are available on board to take control and to operate the shipboard systems and functions.

- *Degree three: Remotely controlled ship without seafarers on board*: The ship is controlled and operated from another location. There are no seafarers on board.

a) UAV/drones for packet delivery for everywhere (Source: Freepik)

b) autonomous/automated guided vehicles for seaports here in Rotterdam; [2].

c) AGV in Amazon warehouses; [3].

d) Autonomous vessels; [1].

Figure 1.
Vehicles on roads, in the air, in seaports, on the sea, in warehouses and all need to communicate with their environments or remote servers [1–3].

- *Degree four: Fully autonomous ship*: The operating system of the ship is able to make decisions and determine actions by itself.

All these types of autonomous vehicles require efficient and highly reliable vehicular communications to exchange data and commands with its neighborhoods, in form of other vehicles, sensors, machines (on wheels), the servers in the remote control centers, etc.

2. Internet of vehicles

The term "Internet of Vehicles" (IoV) refers to the integration of vehicles with the Internet and other communication technologies to enable them to communicate with each other, with infrastructure, and with various services. IoV aims to enhance the safety, efficiency, and overall experience of transportation through the use of connected devices and advanced communication technologies. As illustrated in **Figure 2**, IoV is the evolution of the concept of vehicular communications (initially known as vehicle-to-vehicle or V2V). Its objectives are to cover a wider range of interactions between the vehicles and their environments (neighborhoods and remote servers/control centers) to cover: Vehicles and Personal Services (V&P), vehicle-to-vehicle (V2V), vehicle and roadside (V&R), vehicle and sensor (V&S), and vehicle and infrastructure (V&I or V2I) [4, 5].

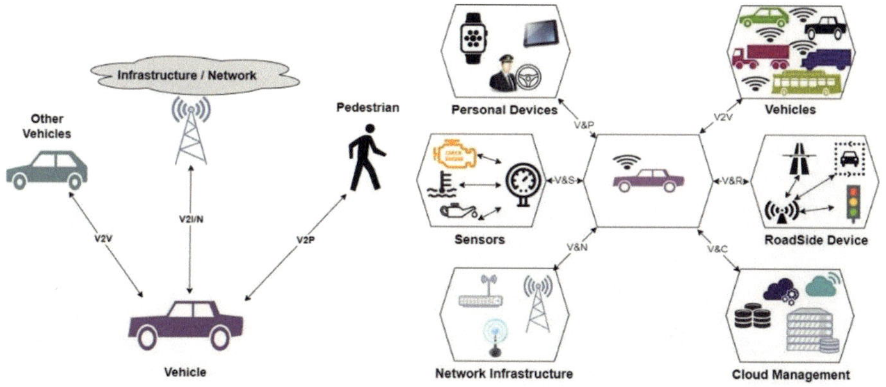

Figure 2.
Evolution of evolution of vehicular communications in form of V2X in LTE/4G (left) toward Internet of Vehicles (IoV) in 5G and Beyond-5G (right).

Generally, the key components of the Internet of Vehicles include:

- *Vehicular communication*: Vehicles equipped with communication devices (such as sensors, GPS, and communication modules) can exchange information with each other and with the surrounding infrastructure. This communication can be used for various purposes, including traffic management, collision avoidance, and cooperative driving.

- *Infrastructure integration*: IoV involves integrating vehicles with the existing transportation infrastructure, such as traffic lights, road signs, and other roadside systems. This integration allows for better traffic management, improved safety, and enhanced overall transportation efficiency.

- *Data exchange and processing*: IoV relies heavily on the exchange of data between vehicles and infrastructure. This data can include information about traffic conditions, road hazards, and the status of other vehicles. Advanced data processing and analysis are essential for making real-time decisions and optimizing transportation systems.

- *Intelligent transportation systems (ITS)*: The notion of IoV is closely related to intelligent transportation systems, which use different technologies to manage and optimize transportation networks. The ITS involves the application of information and communication technologies to transportation infrastructure and vehicles to improve safety, mobility, and environmental sustainability.

- *Safety and collision avoidance*: One of the primary goals of IoV is to improve road safety. Connected vehicles can share information about their speed, location, and other relevant data, enabling systems that can warn drivers about potential collisions or hazardous road conditions.

- *Autonomous vehicles*: IoV plays a crucial role in the development and deployment of autonomous vehicles. Connected vehicles can share data to improve the decision-making process of autonomous systems, enhancing their ability to navigate and respond to dynamic traffic conditions.

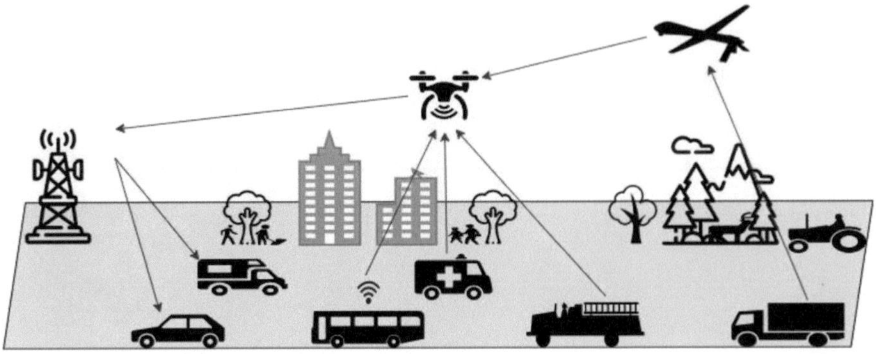

Figure 3.
Vehicular communications covers also unmanned aerial vehicles (UAVs) interacting/communicating with personal cars, busses, trucks, etc.

- *Enhanced driver experience*: IoV can offer various services to enhance the overall driving experience, such as real-time navigation updates, predictive maintenance alerts, and in-vehicle entertainment options.

- *Sensors and connectivity*: IoV relies on sensors, such as GPS, radar, and cameras, installed in vehicles to collect and transmit data. Connectivity is achieved through technologies, such as cellular networks and dedicated short-range communication (DSRC) systems.

- *Data analytics and cloud computing*: The massive amount of data generated by connected vehicles is processed and analyzed advanced technologies, which covers data analytics and cloud computing technologies. This enables real-time decision-making, predictive maintenance, and the development of intelligent transportation systems.

The Internet of Vehicles is part of the broader concept of the Internet of Things (IoT), where everyday objects are connected to the Internet to enable communication and data exchange. The implementation of IoV has the potential to revolutionize transportation by making it safer, more efficient, and more sustainable. The IoV may include also other vehicle types, such as those cited in the previous section, as shown in **Figure 3**. This example includes unmanned aerial vehicles or drones in different forms and types. Either such UAV can be an operational vehicle with specific tasks such as delivery, video surveillance, etc. or it can be a part of communications infrastructure, such as the UAV serving as a base station for the mobile cellular communications [6].

3. Vehicular communications

Vehicular communication refers to the exchange of information between vehicles and between vehicles and roadside infrastructure. This communication is typically facilitated by wireless technology, allowing vehicles to communicate with each other and with the surrounding infrastructure to improve road safety, traffic efficiency, and overall transportation systems. In general, there are two main types of vehicular communication:

- *Vehicle-to-vehicle (V2V) communication*: In V2V communication, vehicles exchange information directly with each other. This can include data about the vehicle's speed, position, acceleration, and other relevant parameters. The goal is to enhance safety by allowing vehicles to be aware of each other's presence and take appropriate actions to avoid collisions. For example, if one vehicle detects a potential collision with another, it can alert the driver or even take autonomous actions to prevent an accident.

- *Vehicle-to-infrastructure (V2I) communication*: V2I communication involves the exchange of information between vehicles and roadside infrastructure such as traffic lights, road signs, and other intelligent transportation systems. This communication can provide drivers with real-time information about traffic conditions, road closures, or other relevant updates. It also allows the infrastructure to communicate with vehicles, optimizing traffic flow and improving overall transportation efficiency.

3.1 Vehicular communications is a complex landscape

Vehicular communications, also known as V2X (Vehicle-to-everything) communication, involve the exchange of information between vehicles and other entities such as infrastructure, pedestrians, and the surrounding environment. Various types of vehicles benefit from vehicular communications, and these technologies play a crucial role in improving road safety, traffic efficiency, and overall transportation systems. The types of vehicles that can benefit from vehicular communications include:

- *Passenger vehicles*: Standard automobiles equipped with communication capabilities can benefit from V2X communication for enhanced safety and improved traffic flow.

- *Commercial vehicles*: Trucks, buses, and other commercial vehicles can use vehicular communications to optimize logistics, improve fuel efficiency, and enhance overall fleet management.

- *Public transport*: Buses and other public transportation vehicles can utilize V2X communication to improve passenger safety, provide real-time transit information, and optimize public transit routes.

- *Emergency vehicles*: Ambulances, fire trucks, and police vehicles can use vehicular communications to prioritize their movement through traffic, improving emergency response times.

- *Cyclists and motorcyclists*: V2X communication can enhance the safety of cyclists and motorcyclists by alerting other vehicles about their presence, especially in blind spots.

- *Pedestrians*: Pedestrians equipped with smart devices or wearables can interact with vehicles equipped with V2X technology, enhancing safety at crosswalks and intersections.

- *Autonomous vehicles*: Self-driving or autonomous vehicles heavily rely on V2X communication to share information with other vehicles, infrastructure, and pedestrians to navigate safely and efficiently.

- *Infrastructure (V2I)*: Roadside infrastructure, such as traffic lights, signs, and sensors, can communicate with vehicles to provide real-time traffic information and signal timing to enhance the overall traffic management.
- *Specialized vehicles*: Vehicles used in specialized contexts, such as construction vehicles or agricultural machinery, can benefit from V2X communication to improve coordination and safety.
- *Electric vehicles (EVs)*: Electric vehicles can use V2X communication to optimize charging schedules, share information about charging station availability, and coordinate with the power grid.
- *Delivery vehicles*: Vehicles involved in package delivery and logistics can use V2X communication to optimize delivery routes, reduce congestion, and improve efficiency.
- *Car-sharing and ride-sharing vehicles*: Vehicles involved in car-sharing and ride-sharing services can benefit from V2X communication to enhance safety and optimize pick-up and drop-off locations.
- *Unmanned aerial vehicles (UAV) /Drones*: UAV can be either an operational vehicle with specific tasks such as delivery, video surveillance, etc., or it can be a part of communications infrastructure, such as the UAV serving as base station for the mobile cellular communications.
- *Automated guided vehicles (AGV)*: These have a large utilization in different industrial domains, such as the AGVs used to transport containers on maritime port arrays.

Vehicular communications are a key enabler for the development of intelligent transportation systems (ITS), providing a networked environment where vehicles and infrastructure work together to improve road safety, traffic flow, and overall transportation efficiency. The integration of V2X technologies contributes to the development of connected and autonomous vehicles and helps build more intelligent and responsive transportation networks.

3.2 Challenges for vehicular communications

While vehicular communication holds great promise for improving road safety and traffic efficiency, there are several challenges that need to be addressed for its successful implementation. Some of the key challenges include:

- *Quality of service (Throughput, Reliability, and Low Latency)*: Vehicular communication systems require low-latency and high-reliability communication to support safety-critical applications. Delays in transmitting and receiving messages could lead to accidents.
- *Security and privacy concerns*: Ensuring the security of vehicular communication systems is crucial to prevent unauthorized access, cyberattacks, and data breaches. Additionally, preserving the privacy of drivers and their data is a significant concern.

- *Scalability*: As the number of connected vehicles increases, the vehicular communication system must be able to scale to handle a growing volume of data and connections without compromising performance.

- *Interoperability*: Achieving interoperability between different manufacturers and models of vehicles is a challenge. Standardization of communication protocols and data formats is essential to ensure seamless communication between diverse vehicles and infrastructure.

- *Spectrum allocation*: The availability and allocation of suitable radio frequency spectrum for vehicular communication is a critical issue. As more devices and systems compete for limited spectrum resources, the interference management and the guarantee of reliable communication become very challenging.

- *Environmental factors*: Vehicular communication systems must contend with various environmental factors, such as interference from buildings, terrain, and other obstacles. Adverse weather conditions, such as heavy rain or snow, can also affect signal quality.

- *Infrastructure deployment*: The deployment of roadside infrastructure, such as communication hubs and sensors, can be logistically challenging and costly. Achieving widespread coverage in both urban and rural areas is essential for the success of vehicular communication systems.

- *Human factor considerations*: Human behavior, including drivers' acceptance of and adherence to vehicular communication systems, is a critical factor. Educating drivers about the benefits and usage of these systems is crucial for their effective implementation.

- *Regulatory and legal issues*: Developing and enforcing regulations related to vehicular communication is a complex task. Legal frameworks must be established to govern issues such as liability, data ownership, and compliance with safety standards.

Addressing these challenges requires collaboration among stakeholders, including automobile manufacturers, government agencies, communication technology providers, and standardization bodies. Continuous research and development efforts are necessary to overcome these obstacles and realize the full potential of vehicular communication systems.

3.3 Technologies and standards for vehicular communications

Several standards have been developed over the years to govern vehicular communications, ensuring interoperability, security, and consistency across different systems. A possible categorization of these standards is illustrated in **Figure 4**, based on a proposition from [7], showing three main categories:

The (traditional) vehicular communications: this class includes some of the first solutions for vehicle-to-vehicle communications, with:

a. *IEEE 802.11p (Wireless Access in Vehicular Environments - WAVE)*: IEEE 802.11p is an amendment to the IEEE 802.11 standard that specifically addresses wireless

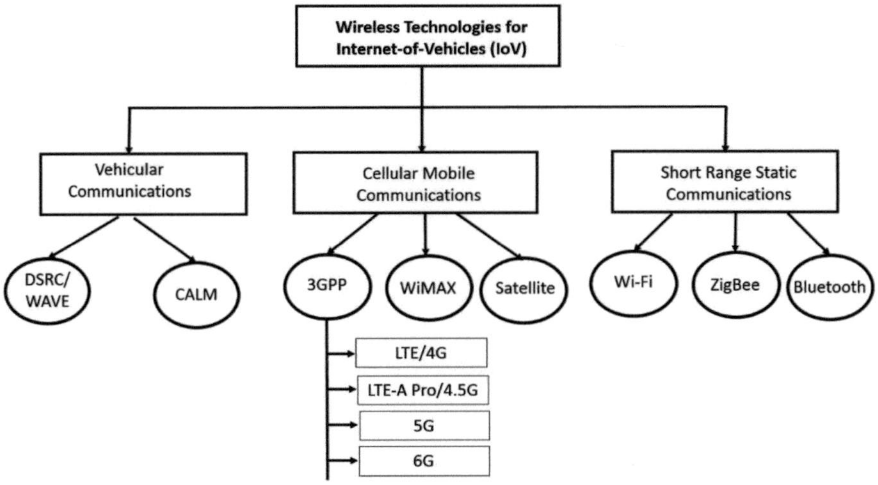

Figure 4.
General classification of wireless communications technologies for vehicular communications and applications, extension based on [7].

communication in vehicular environments. It defines the use of the 5.9 GHz frequency band for dedicated short-range communication (DSRC), supporting V2V and V2I communication. IEEE 802.11bd (Next-Generation V2X - NGV): Under development, IEEE 802.11bd aims to provide advancements beyond IEEE 802.11p with improved performance and new features for next-generation V2X (Vehicle-to-everything) communication. IEEE 1609 Family (Wireless Access in Vehicular Environments - WAVE Standards): The IEEE 1609 family includes a set of standards for WAVE (Wireless Access in Vehicular Environments), covering various aspects of vehicular communication, including networking, security, and message sets.

b. *ISO 24102 (Intelligent Transport Systems - Communication Access for Land Mobiles (CALM) - Facilitation of Traffic Management)*: ISO 24102 provides standards for the communication access for land mobiles (CALM) to facilitate traffic management in intelligent transport systems.

The most dynamic category is the one including the cellular technologies referred sometimes as cellular V2X (C-V2X), such as satellites and WiMAX (IEEE 802.16). However, the most dynamic and innovative solutions are included in standards developed by the third generation partnership project (3GPP) [8]. The vehicular solutions for vehicle communications started with use cases in form of device-to-device (D2D) solution in long-term evolution (LTE) referred as 4G to. This solution evolved in its following generation long- term evolution-advanced pro (LTE-A Pro/4,5G) and reaches its maturity within 5G.

The last category is the short distance wireless communication technologies, which are network protocols in which remote nodes are connected over very short distances. Short-range radio communication can minimize power, volume, heat, and cost. The most known technologies of this category are WIFI (IEEE 802.11), ZigBee, and Bluetooth in both versions the legacy Bluetooth and modern version Bluetooth Low Energy (BLE).

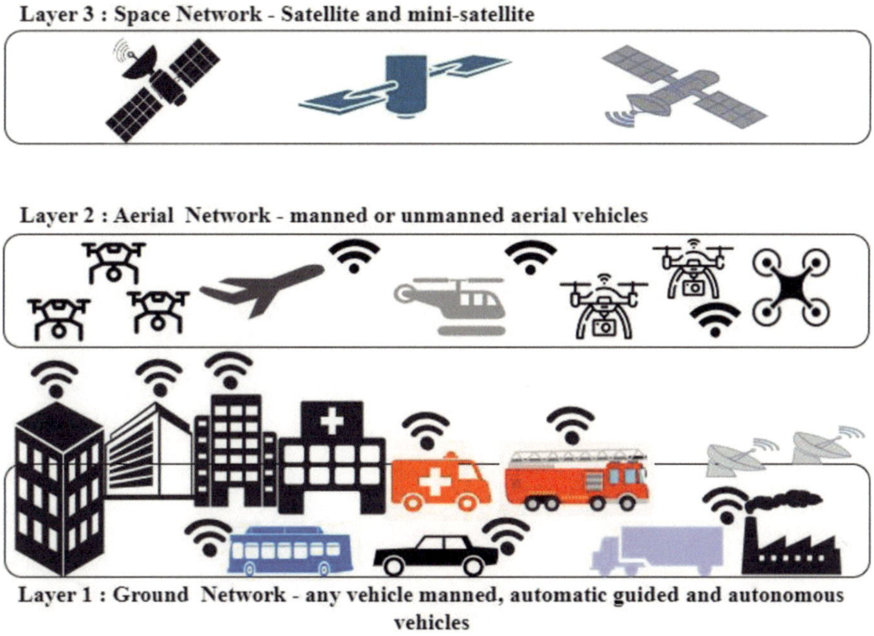

Figure 5.
Vehicular communications and IoV in the heart of development scenarios for the future generation of mobile networks 6G.

These standards play a crucial role in ensuring that different vehicles and infrastructure components can communicate effectively, promoting safety, efficiency, and interoperability in vehicular communication systems. It is important to note that standards may continue to evolve as technology advances and new requirements emerge. It is important to note that standards may vary regionally, and different countries or organizations may adopt specific standards based on their regulatory and technical requirements. Additionally, the field of vehicular communications is dynamic, with ongoing developments and updates to standards to address emerging challenges and technologies. Therefore, the ongoing works on 3GPP standards for the sixth generation of mobile cellular communications networks (6G) are considering more complex use cases and configurations in Internet of Vehicles scenarios [9]. Such future architecture covers three layers that are completing each other and interacting, as shown in **Figure 5**.

4. Conclusions

Vehicles are present everywhere in our daily life, in our industries, and societies. The Internet of Vehicles represents an innovative solution to connect all types of vehicles with the neighborhoods and infrastructures. In the paradigm of Internet of Vehicles, vehicular communications play a key role and must fulfill a long list of requirements. Therefore, different categories of wireless communications have been developed over the time. Standards, especially those issued from 3GPP with its work on the sixth generation of mobile communications (6G), are continuing its adaptability to the new service requirements and challenges to offer an economically viable, highly reliable, and performing adaptive communications solution.

Conflict of interest

The authors declare no conflict of interest.

Author details

Abdelfatteh Haidine
National School of Applied Sciences El Jadida (ENSAJ), University Chouaib Doukkali, El Jadida, Morocco

*Address all correspondence to: haidine.a@ucd.ac.ma

References

[1] Baluco LTD: Maritime Autonomous Surface Ships – Identifying and Covering the Risks; 05.03.2019. Available from: https://www.baluco.com/maritime-autonomous-surface-ships-identifying-and-covering-the-risks/

[2] AGV Rotterdam: Port of Rotterdam: VDL Provided AGVs for the Delta Terminal Rotterdam Which can lift Loads up to 70 Tones. 1 January 2009. Available from: https://www.vdlautomatedvehicles.com/references/port-of-rotterdam

[3] Alfredo PT. 12 Types of Amazon Warehouse Robots. MOBILE ROBOT WORLD NEWS; AGV and AMR Robot Home. Available from: https://www.agvnetwork.com/robots-amazon

[4] Contreras-Castillo J, Zeadally S, Ibáñez JAG. A seven-layered model architecture for Internet of Vehicles. Journal of Information and Telecommunication. 2017;**1**(1):4-22. DOI: 10.1080/24751839.2017.1295601

[5] Seo H, Lee K-D, Yasukawa S, Peng Y, Sartori P. LTE evolution for vehicle-to-everything services. IEEE Communications Magazine. 2016;**54**(6):22-28. DOI: 10.1109/MCOM.2016.7497762

[6] Amponis G, Lagkas T, Zevgara M, Katsikas G, Xirofotos T, Moscholios I, et al. Drones in B5G/6G networks as flying base stations. Drones. 2022;**6**:39. DOI: 10.3390/drones6020039

[7] Kaiwartya O, Abdullah AH, Cao Y, Altameem A, Prasad M, Lin C-T, et al. Internet of vehicles: Motivation, layered architecture, network model, challenges, and future aspects. IEEE Access. 2016;**4**:5356-5373

[8] The 3rd Generation Partnership Project (3GPP): 3GPP – The Mobile Broadband Standard. Available from: https://www.3gpp.org/

[9] Mishra D, Vegni AM, Loscrí V, Natalizio E. Drone networking in the 6G Era: A technology overview. IEEE Communications Standards Magazine. 2021;**5**(4):88-95. DOI: 10.1109/MCOMSTD.0001.2100016

Chapter 2

C-V2X Vision in the Chinese Roadmap: Standardization, Field Tests, and Industrialization

Tao Cui, Lantao Li, Zhaoyu Zhang and Chen Sun

Abstract

Cellular-based V2X (C-V2X) technology promoted by Third Generation Partnership Project (3GPP) is gaining increasing attention globally, after many year-long competition with dedicated short-range communications (DSRC) supported by the Institute of Electrical and Electronics Engineers (IEEE) for vehicular to everything communication. As a rising star, China continuously and actively focuses on and contributes to the C-V2X technology development in this technology marathon. Starting from the standardization progress, a quite few Chinese-specific use cases and communication messages are defined for the complicated urban traffics. Based on these dedicatedly designed higher layer protocols, the annual field tests are progressively conducted to testify the interoperability among chipsets, modules, security certificates, and original equipment manufacturer (OEM). Putting enough efforts on industry standards and tests, China is fast commercializing the C-V2X-based road services, for example, robotaxi and robot-bus in representative cities. Thus, in this chapter, we propose to provide sufficient technology views and a summary to show such advanced Chinese C-V2X philosophy.

Keywords: C-V2X, DSRC, sidelink, message dictionary, interoperability, relay

1. Introduction

Connectivity is always regarded as one of the key facilitators for the evolution of the intelligent transportation and the automobile industry. The actual meaning of wireless and/or wired connectivity for vehicles can be interpreted as a communication network for the scenario of "vehicle-road-cloud" by following certain communication protocols and data exchanging specifications. Specifically, such connectivity of V2X (vehicle-to-everything) is wired and/or wireless connection of intra and inter vehicles between vehicles and the road, and the vehicles and the cloud. For inter-vehicle communication, two major technology directions are widely accepted, that is, C-V2X led by 3GPP and DSRC led by IEEE Both technologies target a global solution with a unified series of short-range communication protocols design. For the application layer specification, regional characteristics are the major contributors for specifying customized standards in order to fulfill the demands of various use cases country by

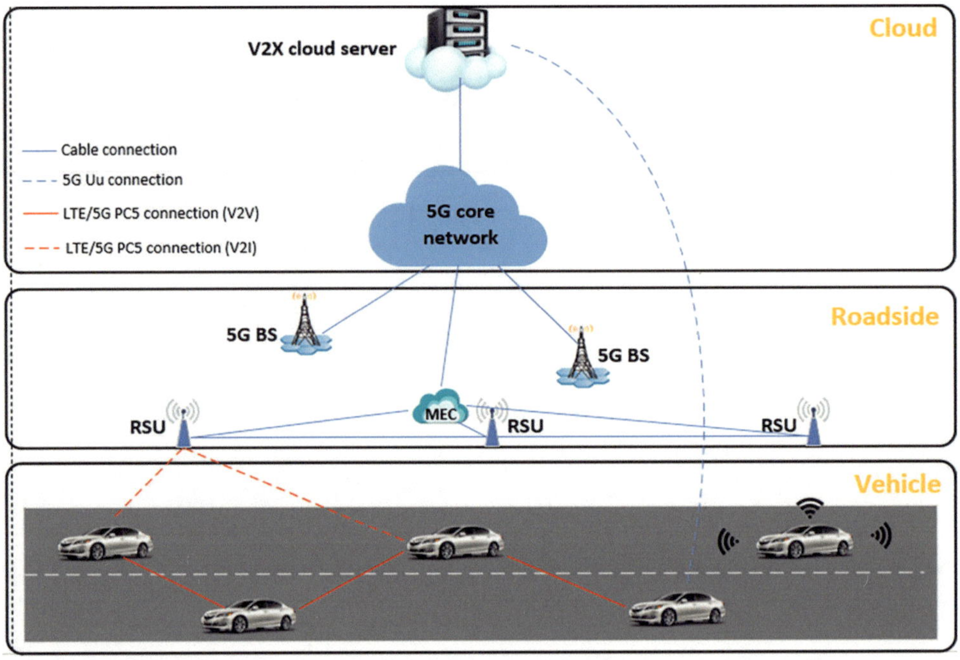

Figure 1.
Vehicle-road-cloud-based V2X system architecture.

country. Though DSRC first version standards came out in 2010, which is far earlier than the freezing of Rel-14 LTE-V2X in 2017, major players in this field started or even implemented cellular-based solutions in many different regions of the world. The main reason behind this is the natural superiority of cellular communication over the IEEE 802.11p on mobility, coverage, and throughput. Hence, China chose the C-V2X technology roadmap in the very beginning, and took the form of a general intelligent transportation system with an architecture of "vehicle-road-cloud" depicted in

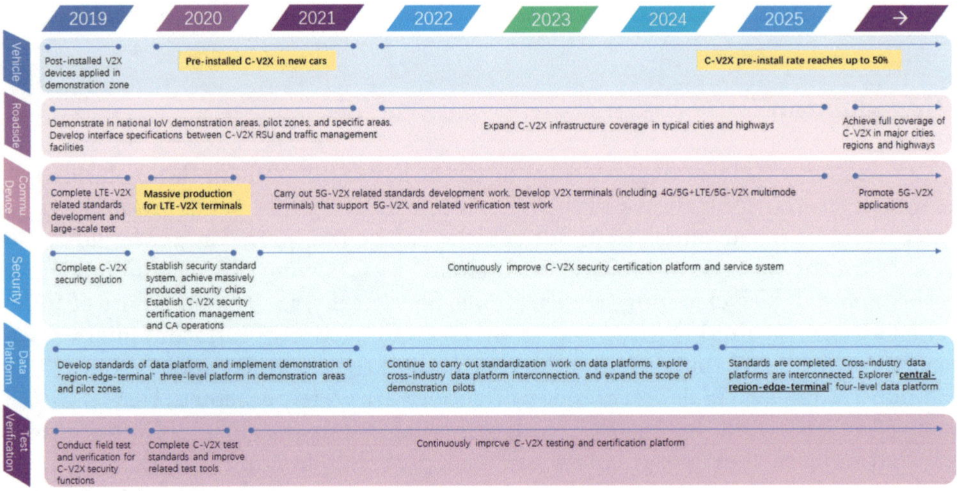

Figure 2.
China C-V2X timeline and milestones.

Figure 1. The hierarchical structure takes full advantage of wireless communication to build up a China-specific intelligent transportation ecosystem. From both the regulation and standardization point of view, China paves a way for a booming C-V2X testification and industrialization in a number of pilot zones. Sections 3 and 4 show the progress and examples in detail (**Figure 2**).

2. Standardization of C-V2X in China

In November 2018, an agreement of C-V2X standards cooperation framework protocol on the reinforcing management of vehicles, intelligent transportation, communication, and transportation management was jointly signed by National Technical Committee 485 on Communications of Standardization Administration of China, National Technical Committee of Auto Standardization (NTCAS), China National Technical Committee and National Technical Committee 576 on Traffic Management of Standardization Administration of China in order to accelerate the C-V2X associated standardization and industrialization.

When it comes to the standardization progress, a list of standards is in **Table 1** across C-V2X general architecture, air interface, safety, network layer, message layer focusing on technical aspects, and test specifications are specified incorporating vehicle, roadside, base station, and core network. Complying with 3GPP specifications on the general framework of sidelink operation, China puts a lot of efforts on the message layer protocol design, whose target is to meet the requirements of different C-V2X use cases from the application point of view. Based on such understanding and

Standards category	Standards	Standards level	Standards organization
Access Protocol	LTE-based general requirements for C-V2X communication system	Industry/national standard	CCSA/TC485
	LTE-based air interface requirements for C-V2X communication system	Industry/national standard	CCSA/TC485
Network Protocol	Cooperative intelligent transportation system-vehicular communication application layer specification and data exchange standard	Association standard	CSAE/C-ITS
	Technical requirements of network layer of LTE-based vehicular communication	Industry standard	CCSA
	Test methods of network layer of LTE-based vehicular communication	Industry standard	CCSA
Message Protocol	Cooperative intelligent transportation system-vehicular communication application layer specification and data exchange standard	Association standard	CSAE/C-ITS
	Technical requirements of message layer of LTE-based vehicular communication	Industry standard	CCSA
	Test methods of message layer of LTE-based vehicular communication	Industry standard	CCSA

Standards category	Standards	Standards level	Standards organization
Security Protocol	Technical requirements of security of LTE-based vehicular communication	Industry standard	CCSA
	Technical requirements of security certification of LTE-based vehicular communication	Industry standard	CCSA
	Test methods of security certification of LTE-based vehicular communication	Industry standard	CCSA
Technical Requirement Regulation	Technical requirements of onboard device supporting sidelink communication of LTE-based vehicular communication	Industry standard	CCSA
	Test methods of onboard device supporting sidelink communication of LTE-based vehicular communication	Industry standard	CCSA
	Technical requirements of roadside device supporting sidelink communication of LTE-based vehicular communication	Industry standard	CCSA
	Test methods of roadside device supporting sidelink communication of LTE-based vehicular communication	Industry standard	CCSA
	Technical requirements of base station equipment of LTE-based vehicular communication	Industry standard	CCSA
	Test methods of base station equipment of LTE-based vehicular communication	Industry standard	CCSA
	Technical requirements of core network equipment of LTE-based vehicular communication	Industry standard	CCSA
	Test methods of core network equipment of LTE-based vehicular communication	Industry standard	CCSA
	Technical requirements of sidelink communication of LTE-based vehicular communication	Association standard	CSAE/C-ITS
	Technical requirements of onboard information exchange of LTE-based V2X sidelink communication	National standard	NTCAS

Table 1.
LTE-based C-V2X standards in China.

envision, standards of cooperative intelligent transportation system: Application layer specification and data exchange standard (Phase I and Phase II) were accomplished by C-SAE and C-ITS in 2017 and 2020, respectively. There are 34 use cases specified in total, with dedicatedly designed messages for each of them, except for the basic message set constituting of five messages, that is, basic safety message (BSM), signal phase and timing message (SPAT), road safety message (RSM), map data, and road side information message (RSI). Compared to the data dictionary defined by the Society

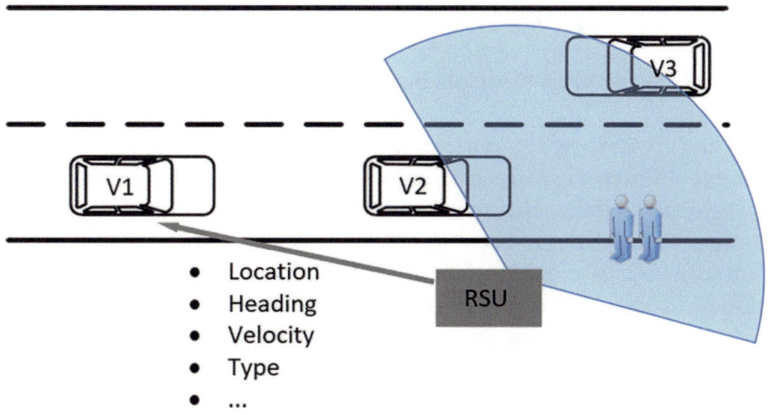

Figure 3.
RSU-aided sensor data sharing to avoid blind zones.

of Automotive Engineers (SAE) J2735 [1], the Chinese standards are compatible for a general road and vehicle attribute on the basis of the data frame and data element.

In a typical use case of sensor sharing, RSM and sensor sharing message (SSM), which are not defined by the US and EU are utilized for enlarging the host vehicle sensing range, especially at its blind zone, with the help of sensor data collection from roadside sensors and other vehicles, respectively. An example of roadside unit (RSU)-aided sensor sharing scenario is given in **Figure 3** while the detected object is notified along with the attributes of type, location, and status. It is well noted that such information-sharing mechanism only prefers the exchange of formatted data after being processed to the point-cloud-based raw sensing data due to the constraint of C-V2X bandwidth (20MHz for LTE and 40MHz TBD for new radio (NR)). The sharing of raw data has opportunities to become possible in the future once the operation frequency band goes up to the mmWave with comparable bandwidth to

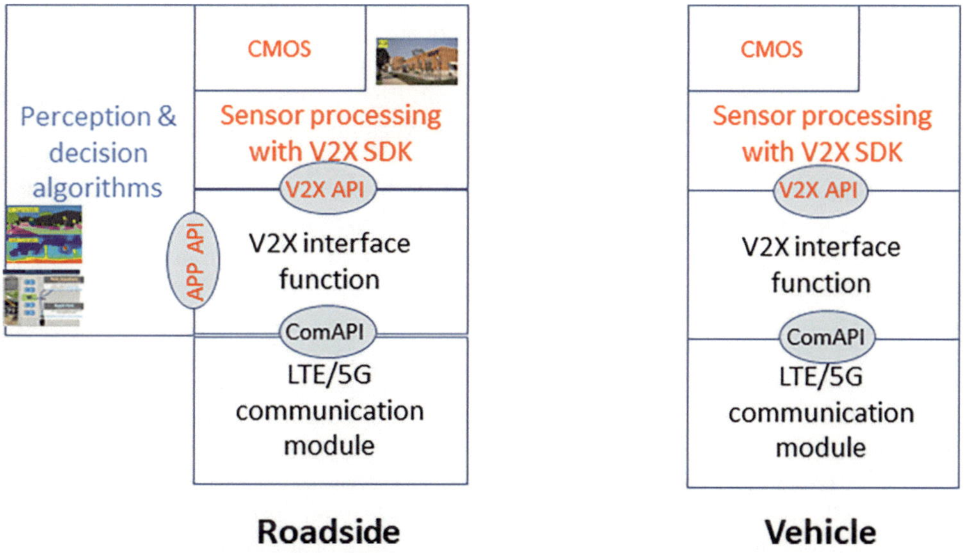

Figure 4.
Sensor data sharing reference model.

the Uu link, which was approved as a work item for one of the new features for 3GPP Rel-18 [2]. The undergoing discussion targets a frequency range-2 (FR2), such as solutions on the Uu link, with multiple beams operation for stable and robust sidelink communications.

For better clarification of the sensing data exchanging principle, the following reference model (**Figure 4**) shows different entities and corresponding interfaces for supporting the sensing data routing in the system. In a normal working flow, the sensor module outputs Red-Green-Blue (RGB) image data via direct inputs from CMOS. Imported into the V2X interface function via V2X API, post-processing is facilitated to turn the image data into the given format that RSU recognizes, that is, object type, status, location, etc. Through the communication API, the sensing data contained in RSM or SSM are sent out over LTE or 5G communication module. If the application perception and decision engine request sensing data from other nodes of the traffic, the data flow will be imported to the engine over the communication module and application API at the receiver. It is well noted that the exchanging data could be either decision-level or raw data-level depending on the sensor fusion algorithms and system tolerated latency at the engine.

3. C-V2X field test

In China, the C-V2X large-scale interoperability test is recognized as the most influential demo show and test activity in the C-V2X field, which is held by IMT-2020 (5G) promotion group and aims at testing and verifying the various applications of cooperative vehicle infrastructure system through the real application scenarios. At this point, the C-V2X working group under IMT-2020 (5G) promotion group plays a vital role in organizing the activities, including the C-V2X research, on core technology verification and the promotion of the C-V2X industry based upon China's intellectual property principle. With the advancement of C-V2X technology and standard, the C-V2X large-scale interoperability test has also extended its application scenarios, from driving safety to driving efficiency in the annual test event. The test results not only contribute to the improvement of products but also become samples to provide a new basis for the construction of the national demonstration area of the V2X. In Nov. 2018, the activity of three layers interoperability V2X application demonstration was held in Shanghai for its world first time, showcasing the interoperability for the interconnection cross chip modules, terminals, and vehicles. 3GPP Rel-14 LTE-V2X specifications were adopted with regulated 20MHz from 5905Mhz to 5925Mhz as a physical layer sidelink principle. Meanwhile, the Chinese-specific message layer protocol of cooperative intelligent transportation system – vehicle communication application layer specification and data exchange standard (Phase I) was taken as a message dictionary in the activity. On top of "Three Layer" event, the C-V2X working group held a "Four Layer" demonstration activity by emphasizing communication security scenarios in Oct. 2019. With such joint verification, it paves a way for China C-V2X in its commercialization process. To be further closer to the actual traffic scenarios, the third term event called "New Four Layer" was promoted in Oct. 2020 by considering a high-precision map.

In Nov. 2021, Sony participated in the C-V2X large-scale interoperability test as one of the fourth-term event organized by the China Academy of Information and Communications Technology (CAICT). The test field is located in Wuhan economic development district, a blocked T-junction road area with 200 OBUs

Figure 5.
BEV of the 2021 Wuhan C-V2X large-scale interoperability test field.

arranged alongside to simulate realistic C-V2X packet transmission in a crowded traffic area seen in **Figure 5**. A series of required tests were conducted for different application scenarios, the test result shows that the packet reception rate remains highly stable (e.g., 95% or above) within the transmission range of 50 meters, and gradually declines to 80% or worse with transmission range growing to 400 meters. As observed in the bird's eye view figure that the road is curved at a certain level, and trees and bushes in the middle of the road happen to be a natural fence for long transmission between sender and receiver along the path of B to D, and the rough relief terrain between B and C also leads to typical NLOS transmission along the path of B to C. The worst test result shows that the data packet reception rate of 70% or even 60% may be achieved based on the current Rel-14 C-V2X chipset OBU with a transmission range from 300 meters to 400 meters (**Figure 6**).

To address the signal attenuation issue caused by obstacles, a Veins simulation platform is built up, which is based on simulation of urban mobility (SUMO) for mobility and objective modular network testbed in C++ (OMNeT++) for communication. The reason behind choosing Veins is that, besides the signal path loss model provided, the obstacle shadowing model [3] and vehicle obstacle shadowing model [4] added in the latest version of Veins make it easy to focus on how to define better-optimized relay selection algorithm in the upper layers to enhance V2X communication coverage. A very typical urban intersection area near Berlin is selected with around 30 vehicles near or approaching the intersection center, the vertical road has six lanes, and the horizontal road has only two lanes, thus different densities of vehicles are applied to different roads. In the no relay simulation, vehicles alongside the horizontal road achieved the data packet reception rate at 29.5%, while vehicles on the

Figure 6.
Sony's participation in the C-V2X large-scale interoperability test.

vertical road at 75.1%, and vehicles within 50 meters range to the intersection center could achieve 95% above data packet reception if dividing simulation periods and focus on packet reception rate variance corresponding to location variance. This is a natural consequence since the LOS transmission path dominates the scenario when either sender or receiver is at the intersection, while NLOS is typical when both two sides are far from the center and not in the same section of road. The square area with 100 meters side length is specifically selected as the relay selection area, all vehicles within this area could switch their relay function on, and the maximum relay nodes number is applied to minimize redundancy of relayed transmission. Multiple methods could be used for relay node candidate confirmation, in our simulation, we use an announcement message to identify the existence of relay nodes to eliminate potential relay redundancy, the candidate node must listen for a period before switching on the relay function. With the relay switched on in simulation, the performance increased to 91.3% for horizontal direction vehicles and 98.9% for vertical direction vehicles, at the cost of 1.99 times more transmission budget for relaying data packets (**Figure** 7) [5].

It is a popular design in the C-V2X demonstration area that RSU will be deployed at such areas for information collection, information refinement, and relay function, however requiring substantial infrastructure costs, the vehicle-based relay may provide another route. Though such relay candidate selection, the area seems easy to find, and the various real road conditions introduce a wide variety of scenario-specific difficulties in accurately identifying the area in details. However, with the development of network data analytics function (NWDAF) and other artificial intelligence (AI) topics, such as federated learning introduced into 3GPP, the exposure of V2X communication characteristics from UE to gNodeB and core network may provide a platform for data analysis, relay on-request area, and relay candidate selection area determination. Similar to how the base station manages the area of the cell, the high-definition map can also be divided into different V2X communication blocks with the intersection area, the bend area, and semi-open location close to obstacles as relay node selection areas. The mechanism to define transmission block and intersection should be based on the quality of service (QoS) of packet transmission, the QoS

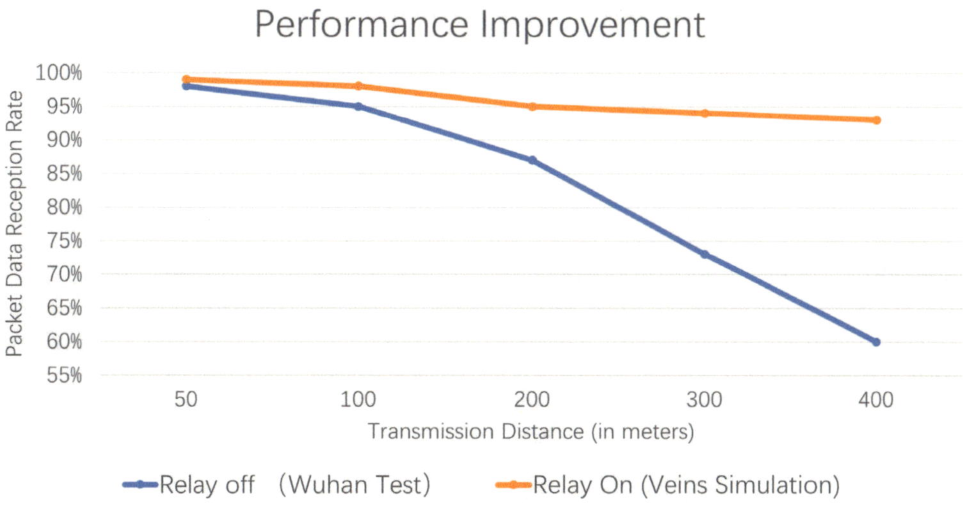

Figure 7.
Data reception rate performance improvement.

of packet transmission within one V2X communication block shall remain aligned and stable, while Tx UE and Rx UE belong to different communication blocks that differ largely with location difference. The relay node selection area shall be bridging multiple communication blocks, while the UEs inside the relay node selection area can establish stable short-range transmission link with UE in those surrounding multiple communication blocks. The base station may take the role of data collector to identify short-range data transmission quality characteristics with the locations of both sender and receiver, reporting the statistics to the core network, enabling NWDAF to analyze the V2X transmission quality in relation to different areas with the help of location management function (LMF), to decide V2X communication blocks and relay node selection areas. Vehicle type UE entering a relay node selection area can be validated by the base station or core network functions as relay nodes dynamically to minimize the non-line-of-sight (NLOS) transmission from obstacle (e.g., buildings and vehicles) shadow effects.

The solution introduced above is just one example of how the C-V2X large-scale interoperability test exposes technique issues and triggers corresponding solutions. Sony China will continue its participation in the coming years and make attempts to implement such solution in the module for validation under a more realistic C-V2X application environment.

4. Industrialization of C-V2X in China

C-V2X technology has been developing for several years. Manufacturers continue to invest in integrated circuit (IC) design, on-vehicle devices, and roadside devices. Countries around the globe conducted test fields to verify the technology and performance of C-V2X. In the United States, the University of Michigan cooperated with Verizon to build a test bed that deploys video cameras with a 5G communication module to develop 5G solutions for autonomous and connected vehicles [6]. In Europe, a trinational test bed of C-V2X was built in 2019 in the border region of Germany, France, and Luxembourg [7]. In 2020, Softbank and Subaru testified use cases related to the safety of autonomous driving with C-V2X [8]. Comparing with the United States, Europe, and Japan, China invests and sets up more test fields with more cross-industry companies participating in deploying C-V2X in the country.

4.1 C-V2X deployment supported by the Chinese administration

In April 2021, the Chinese government announced the first six cities as pilot city for intelligent and connected vehicles (ICV). The six cities are Beijing, Shanghai, Guangzhou, Wuhan, Changsha, and Wuxi. The main targets for the pilot city include building intelligent infrastructure, enhancing network connectivity, and explore the function of "vehicle-road-cloud" platform, which is a unique system architecture for ICV in China.

In December 2021, another 10 cities were announced by the Chinese Government as the second pilot cities for ICV, including Chongqing, Shenzhen, Xiamen, Nanjing, Jinan, Hefei, Chengdu, Cangzhou, Wuhu, and Zibo. Specific goals were set for these cities. For example, the target of Chongqing is implementing ICV technologies into public transportation to improve the avenue of the tourism industry; the target of Nanjing is to explore the functions of ICV for the customer; and the target of Jinan is to combine the ICV technologies with the agricultural logistics industry.

From 2015, China started to build pilot zone in different cities as test zones for ICV, as shown in **Figure 8**. The purpose of the pilot zone includes testing key technology for ICV, demo platform for R&D testing, experimental verification, and providing standards and legislation for the ICV industry. Among all the pilot zones in China, the Shanghai pilot zone and the National pilot zone (Jingji) are the two very large ones.

The Shanghai pilot zone started in 2015, and it is the first and the most advanced national-level ICV test zone in China. It contains a closed test zone, open testing roads, and an urban roads demo zone. The closed test zone has 15 km test roads, which can provide a test for 200 vehicles. 42 RSUs are deployed in the closed test zone, which can provide DSRC, long-term evolution-based vehicular (LTE-V), and WiFi connection. The open testing roads is 73 km in length along with 182 RSUs, which can provide a test for 1000 vehicles. The urban roads demo zone is an area of

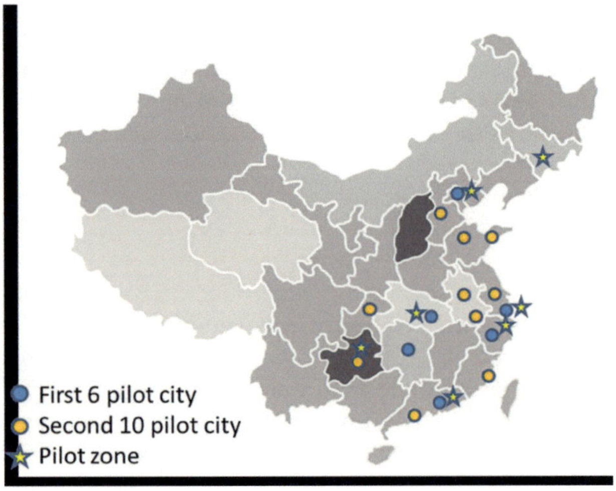

Figure 8.
The pilot cities and pilot zone in China.

100 km^2 that can provide a test for 5000 vehicles at the same time. It can provide more than 200 test scenarios, including an autonomous driving test (lane changing, pedestrian detection), efficiency test (auto valet parking (AVP), platooning, and traffic control), and C-V2X test.

Another National pilot zone (Jingji) supported by Ministry of Industry and Information Technology (MIIT), Ministry of Transportation (MIT), and the local government was established in 2016 in Beijing. It can provide a test for 10000 vehicles with 1000 km test roads with different types of roads (highway, urban road, and country road). It has the first test road specifically for vehicles with the V2X function. It also has a test area with different road scenarios, such as railroad, N road, S curve, and five forks. The Beijing pilot zone publishes an annual report from 2018. In 2021, there were 124 total qualified autonomous vehicles for open road testing in the pilot zone with an accumulated 251 million kilometers of test mileage. From 2018 to 2020, one of the research targets of the pilot zone is a perception of autonomous driving with C-V2X. In its annual reports, the costs of sensors used for perception decreased due to the C-V2X technology and autonomous driving technology upgrade. In 2018, most of the autonomous driving testing vehicles were deployed with 128-beam or 64-beam Velodyne light detection and ranging (LiDAR). The price of 64-beam Velodyna LiDAR was more than 500K RMB. In 2021, most autonomous vehicles were deployed with 40-beam LiDAR. The total cost for one autonomous driving vehicle costs only 480K RMB.

4.2 C-V2X development of industry in China

As the Chinese government accelerates the massive deployment of C-V2X, enterprises in China are leading the market not only by advancing research but also by making it true in the market.

Early in 2017, Huawei, China Mobile, and Shanghai Automotive Industry Corporation (SAIC) demonstrated the world's first 5G-based remote driving with a consumer vehicle, as shown in **Figure 9** [9]. In the demo, Huawei provided the 5G wireless solution, SAIC provided the smart concept car, while China mobile provided

Figure 9.
Remote driving demo by Huawei.

the 5G connectivity. The driver was located 30 kilometers away from the car. 4 HD video cameras were installed on the car, which can provide a 240-degree view of the vehicle's surroundings. The control signals were also transmitted over a 5G network. Huawei said the video streaming was flawless and the end-to-end latency of the control signal was less than 10ms. Two years after, in 2017, Huawei brought a 5G vehicle module MH5000, which was the industry's first 5G vehicle module. It integrates 5G and C-V2X technology. The module has been tested in the 2019 Shanghai C-V2X application demonstration [10].

Pony.ai, a leading autonomous driving technology company in China, has been focusing on the robotaxi service since 2016. It received approval to run autonomous robotaxi services in the Beijing pilot zone in November 2021 [11]. In the same year, it received a permit to conduct public road tests in Shenzhen [12]. Besides the robotaxi service, Pony.ai launched its autonomous trucking tests on the open highway in Beijing. Its autonomous trucking business had already accumulated 50,000 km of commercial operation and delivered over 16,400 tons of goods by the end of 2021.

Apollo is a project of autonomous driving developed by Baidu in 2017. Its consortium is more than 100 companies, which aims to be the world leader in the driverless vehicle industry to the intelligent traffic system. Baidu Apollo launched the world's first V2X platform that enables L4 autonomous driving on open public roads using roadside sensing in May 2021 [13]. The aim of this platform is to achieve autonomous driving through vehicle-road-cloud coordination. Baidu has achieved a cumulative mileage of 2019230 km over a three-year period in Beijing pilot zone, accounting for 91.23% of Beijing's total testing mileage in 2018, 2019, and 2020. It operated a fleet of 55 autonomous vehicles on the roads of Beijing, accounting for 75% of the overall number of vehicles that conducted autonomous driving tests in the capital of

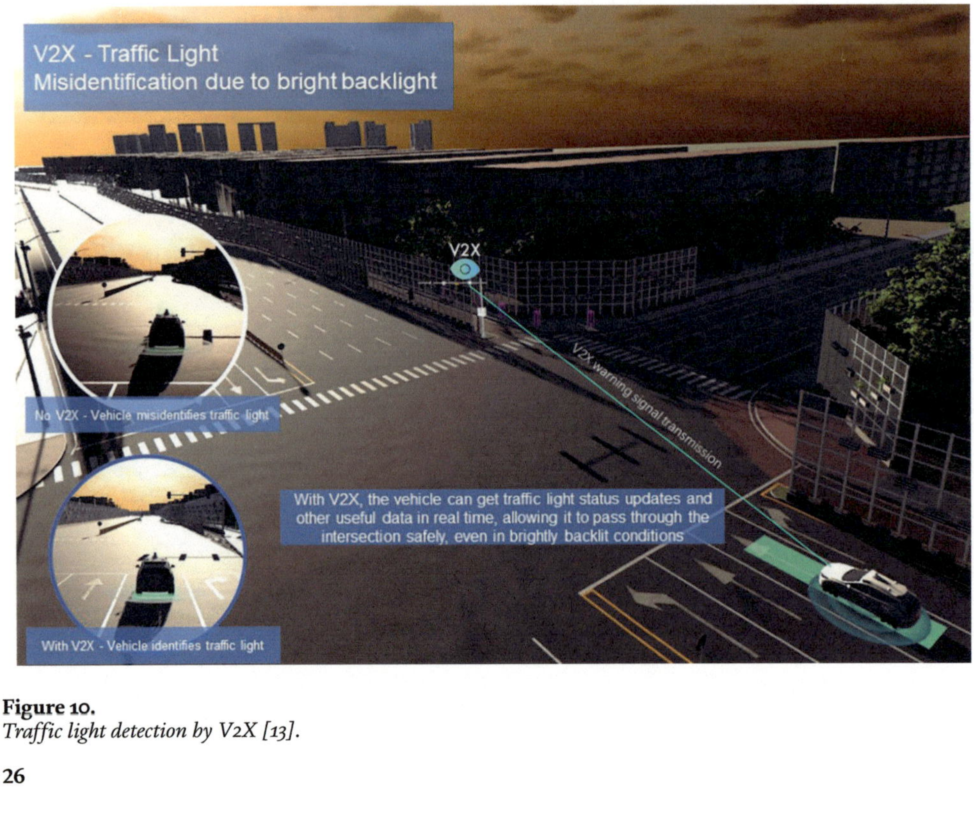

Figure 10.
Traffic light detection by V2X [13].

China. Its robotaxi service began in Beijing in 2020. From October 10 to December 31, 2020, a total of 15006 users used the taxi service through their mobile devices. Chongqing, China, and Baidu provide an intelligent traffic system, which decreases wait time at the intersection by 6.31% and mean time of parking by 2.6 minutes. In order to improve the management of the intelligent traffic system, Baidu also provides an HD-map service that can capture road information, such as lanes, traffic signs, roadside devices, and facilities, with sub-meter level positioning error. The vision of Baidu's V2X business is to enable level-4 autonomous driving on public roads using cooperative sensing by both roadside and on-vehicle sensors. Baidu has tested its cooperative sensing technology with C-V2X communication at multiple intersections in Beijing, Guangzhou, and Cangzhou. One of the test scenario is that, when vehicles turn or change lanes at intersections, roadside sensors can detect road conditions and transmit data to cars in real time. Another scenario that reliable safety can be provided via Baidu's technology is the limitation of on-vehicle perception in severe weather conditions. For example, if the traffic light has a bright backlight, vehicles with only self-sensing system may not correctly identify the traffic signal. However, with V2X, the vehicle can get the traffic light status in real time, allowing it to pass through safely, as shown in **Figure 10**.

Another world's leading tech giant, Didi Chuxing, launched its robotaxi service in Shanghai within a designated public area in 2020. Key C-V2X devices were deployed on the road to minimize the blind spots and enhance the coordination among the autonomous taxi fleet in order to guarantee the safety of the passenger.

5. Conclusion and perspectives

This chapter gives a general view of China-specific C-V2X developing roadmap in terms of standardization, field test, and industrialization. The paradigm of China's ICV is through architecting localized solutions of cyber-physical systems by combining the characteristics of automation and connection, which underlines the concept of "human-vehicle-road-cloud." The hierarchical framework fully utilizes the properties of complex and driving behavior-led traffic in China, and paves the way for ensuring ICV security and industry security under all-dimensional support of the government. As such, there is a reasonable prospect that advanced autonomous driving technology aided by C-V2X is on the way to ease people's lives in a superfast manner.

Author details

Tao Cui*, Lantao Li, Zhaoyu Zhang and Chen Sun
Sony (China) Ltd., Beijing, China

*Address all correspondence to: tao.cui@sony.com

References

[1] Hedges C, Perry F. Overview and use of SAE J2735 message sets for commercial vehicles (No. 2008-01-2650). SAE Technical Paper. 2008

[2] 3GPP RAN Plenary #94e meeting, RP-213678

[3] Sommer C, Eckhoff D, Dressler F. IVC in cities: Signal attenuation by buildings and how parked cars can improve the situation. IEEE Transactions on Mobile Computing. 2014;**13**(8):1733-1745. DOI: 10.1109/TMC.2013.80

[4] Sommer C, Joerer S, Segata M, Tonguz OK, Cigno RL, Dressler F. How shadowing hurts vehicular communications and how dynamic beaconing can help. IEEE Transactions on Mobile Computing. 2015;**14**(7):1411-1421

[5] Cui T, Sun C, Li L.T. Information relaying methods in VANET: Algorithms, standards and tests, submitted to Data Dissemination in Vehicular Networks, Applied Sciences, 2022

[6] Verizon 5G Ultra Wideband network now live at University of Michigan's test bed for driverless cars | About Verizon, Verizon, Available from: https://verizon.com/about/news/verizon-5g-ultra-wideband-university-michigan [Accessed: 04 August 2022]

[7] Vedecom French-German-Luxembourgish cross-border test bed inauguration: VEDECOM presents the TRIICA project [Internet]. Available from: https://vedecom.fr/inauguration-du-site-de-test-transfrontalier-france-allemagne-luxembourg-vedecom-presente-le-projet-triica/?lang=en [Accessed: 04 August 2022]

[8] SoftBank Corp and Subaru Corporation Verify Use Cases Related to Automated Driving | About Us | SoftBank [Internet]. Available from: https://www.softbank.jp/en/corp/news/press/sbkk/2020/20201124_01/ [Accessed: 04 August 2022]

[9] Huawei Demonstrates 5G-based Remote Driving with China Mobile and SAIC Motor – Huawei Press Center [Internet] Available from: https://www.huawei.com/en/news/2017/6/5G-based-Remote-Driving [Accessed: 04 August 2022]

[10] Gearbest. What is Huawei 5G+C-V2X vehicle communication technology? | GearBest Blog [Internet] Available from: https://www.gearbest.com/blog/how-to/what-is-huawei-5gc-v2x-vehicle-communication-technology-6692 [Accessed: 04 August 2022]

[11] Businesswire.com. 2021. Pony.ai Receives Approval for Paid Autonomous Robotaxi Services in Beijing | Business Wire [Internet] Available from: https://www.businesswire.com/news/home/20211129005790/en/Pony.ai-Receives-Approval-for-Paid-Autonomous-Robotaxi-Services-in-Beijing [Accessed: 04 August 2022]

[12] Businesswire.com. 2021. Pony.ai Begins Autonomous Driving Public Road Testing in Shenzhen | Business Wire. [Internet] Available from: https://www.businesswire.com/news/home/20211210005555/en/Pony.ai-Begins-Autonomous-Driving-Public-Road-Testing-in-Shenzhen. [Accessed: 04 August 2022]

[13] Stone T. Baidu launches V2X platform to guide CAVs using roadside infrastructure | Traffic Technology Today [Internet] Traffic Technology

Today. 2021. Available from: https://www.traffictechnologytoday.com/news/autonomous-vehicles/baidu-launches-v2x-platform-to-guide-cavs-using-roadside-infrastructure.html [Accessed: 04 August 2022]

Chapter 3

Mapping the Applications of Vehicular Communications in the Context of Smart Maritime Ports

Abdelfatteh Haidine, Ayoub El Idrissi, Abdelmoula Ait-Allal, Aziz Dahbi and Abdelhak Aqqal

Abstract

The maritime transport networks play a critical and major role in an increasingly globalized world economy. Within these networks, the maritime ports play the role of hubs. Any disturbances in these hubs will negatively affect the worldwide economy. Therefore, economy players are transforming the ports through an evolutionary process to become smart maritime ports. These smart ports are built through an ensemble of smart domains that adopt sensing, data transmission, and data intelligence to support intelligent decision-making processes. Examples of such smart domains include smart grid/microgrids, smart container management, and smart/automatized terminal operations. In each of these domains, optimal decisions must be met to optimize the use of resources, increase the economy efficiency of the ports, and increase the safety and security for assets, goods, and people. In smart maritime port environment, vehicular applications are adopted everywhere, such as automated guided vehicles to transport containers, unmanned aerial vehicles for different port operations, etc. In this work, we discuss some concrete examples of these vehicular applications in the smart port environment and suggest the adequate and optimal vehicular communication technologies to be deployed to support a reliable data transmission for these applications.

Keywords: smart maritime ports, vehicular communications, unmanned aerial vehicles, intelligent transport system, mobile communications

1. Introduction

In today's maritime world, numerous pressures are exercising on ports; so many things have been changed deeply. Therefore, ports worldwide face challenges every day. Nowadays, we assist to more goods and services that are being shipped, resulting in ship's traffic that is increasing permanently [1]. Ports are crowding because of the growth of population in the world, accentuated by the globalization and the growing maritime industry. So far, the maritime world has been impacted by the global trade, sustainability, and digitalization.

IntechOpen

A simplified configuration of maritime transport networks is depicted in **Figure 1**, showing that such a network can be modeled as graph network G = (V, E), where the set E constitutes the maritime routes, which must be secure, cost-efficient, short-delay, and reliable. The set of nodes (or vertices) is constituted by the world maritime ports/seaports. These nodes must also be secure/safe, cost-efficient, short-delay, reliable, environment friendly, and energy-efficient. Any disturbances in these nodes, that is, ports, will result in dramatic scenes and negative impact on the economy, nationally and as well as internationally in more globalized economy/industries. Such a concrete example is depicted in **Figure 1** (bottom) showing an increasing queuing of jam in front of maritime ports. More concretely, port authority recorded shattered 73 container ships stuck waiting off California port [2].

All these maritime challenges and so more could be reduced or even deleted if we focus on the solutions proposed by the new technologies that convert all the drawbacks into opportunities. Nowadays, and in a competitive world, being attractive or even exist needs to innovate with more intelligence and by using the digital way to care more for the future of the industry. Thus, the maritime ports should follow this evolution.

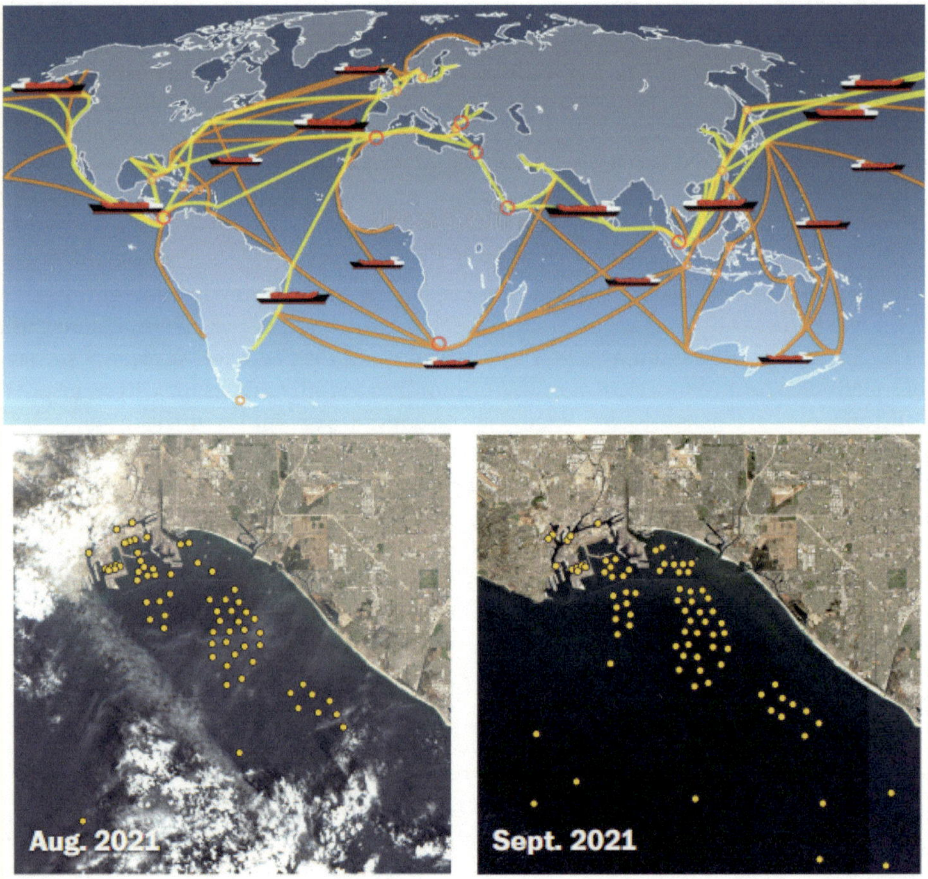

Figure 1.
Simplified presentation of maritime transport network (top, Credits: Dreamstime) and Jam of containerships in front of Los Angeles port resulting in a long delay of containers and goods delivery (bottom, Credits: Washington Post).

According to this context, and including the intelligence to ports, a new concept has emerged, which is called smart port intended to care for the maritime environment in order to be more effective and performing. Smart ports lead to provide more with less. This means saving time, resources, and money. Furthermore, smart ports is aiming to achieve a better connection for stakeholders using the automated ships, to massively reduce pollution and congestions, while increasing time/economic effectiveness and safety.

Smart ports' solutions and strategies can be integrated to solve most of the maritime challenges. The use of new technologies and intelligence is a way to transform the traditional practices of ports and their weaknesses to smart ports with more benefits. Smart technologies, such as big data, artificial intelligence (AI), Internet of Things (IoT), blockchain technology, and fifth-generation (5G) connection, are means to change the maritime industry. They are used as a decision-making system and a way to monitor processes with developing more security and safety in the working environment.

In this chapter, we discuss the main components for the realization of smart maritime ports; then, we focus more on vehicular applications. These applications should lay on a strong performing and reliable communication layer. Therefore, we analyze different possible transmission technologies for vehicular communications through a qualitative evaluation of different performance criteria, which are more relevant for smart maritime port environment.

2. Smart maritime ports and IoT model

2.1 Smart ports: definitions and objectives

Smart is a new concept adapted to the technology setting to follow the modern world. It is a word that contains digital information, computer development, and a way to innovate. According to [3], smartness is defined as the application of automatic computing principles such as self-configuration, self-protection, self-healing, and self-optimization. Smartness has found its way into many industries, especially in urban areas, as a solution to a number of problems such as pollution, increasing cost, problem of congestion, and preservation of historical sites.

Smart ports provide digital port solution consultancy for new and existing ports. Smart ports combine new technologies, automation, and digitalization to optimize the port operations. The use of smart ports is a way to boost digital transformation and make it effective in practice; also, it gathers technology, innovation, and digital tools. Smartness relies on information and digital technologies to become more efficient, sustainable, and collaborative. It encompasses different types of technologies such as big data, artificial intelligence (AI), Internet of Things (IoT), blockchain, and 5G.

Finally, a smart port is a port facility using information and communication technologies (ICT) to increase efficiency and ensure the safety of transshipment operations.

The objective of smart ports is to provide solutions to four key issues: economic issues, ecological issues, energy issues, and social issues. Today, many ports seek to use technology to achieve these objectives and to become a sustainable port responding to economic, ecological, and citizen issues:

- Concerning the economic aspect, the objective of the smart port will be to reduce the costs linked to the increase in flows and goods; many smart ports use artificial intelligence. For instance, AI can be used to predict the arrival of ships or to know the activities of each in real time.

- As for the ecological aspect, and to meet this challenge, smart ports seek above all to reduce their greenhouse gas emission, which also increases the air quality of their city.

- Smart ports aim to address the challenge of energy in two main ways. The first is through the development and use of renewable energy sources, such as solar, wind, and wave power. The second is by reducing overall energy consumption in the ports through the use of energy-efficient technologies and processes.

- From the perspective of the social aspect, ports generally adopt two solutions: education and participatory democracy.

2.2 Smart domains in smart maritime port environment

The evolution of the traditional port into a smart maritime port is complex and involves various research activities. Therefore, the smart maritime port can be seen as a set of (also known as "vertical" domains) smart domains, as depicted in **Figure 2**. In the following, we describe some of these smart domains.

Smart grid: Smart grid technology represents the next major advance in power grid management. With the help of various technological means, it allows an intelligent control of energy flows, enabling an adequate balance between energy supply and demand, resulting in savings for both consumers and producers. Smart grids also play an important role in the development of renewable energies. However, the high implementation costs and certain data protection issues remain major challenges against this technology.

The major applications and roles of the smart grid can be summarized as follows:

- Balance: It allows for better management of the energy produced and the energy delivered.

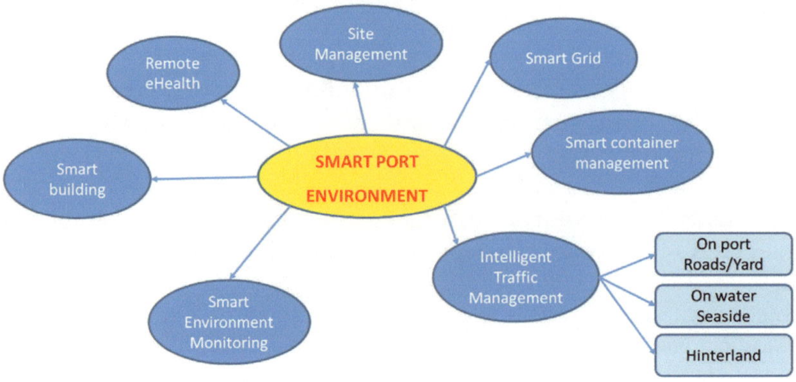

Figure 2.
The smart port environment is constituted of a set of smart domains.

- Security: The smart grid improves the stability of the electrical network, reducing the risk of overloads or outages, which can have negative impacts on both individuals and industries.

- Ecology: The smart grid allows a better use of renewable energies in the global energy consumption and, thus, avoids very important losses.

- Economy: A more stable network management allows a drastic reduction of production and consumption costs and lower losses.

Intelligent transport system (ITS): The state-of-the-art solution for the management of truck flows is based on an innovative concept of an intelligent parking lot, which makes it possible to put an end to port congestion and to fluidify the adjacent road traffic.

In this context, having an access permit is no longer sufficient for gaining access to the port. Access to the port is subject to an appointment with the operator through its online software system connected to automated gates and cameras; trucks accessing the port have the option of using the physical parking lot or the virtual online parking schedule. They will only be allowed into the parking lot after a mission has been created and the port operator has confirmed the scheduling of an imminent operation.

The Internet of Things (IoT) and artificial intelligence (AI) are enabling a new class of intelligent transportation systems (ITS) for road, air, rail, and sea. These solutions connect vehicles, traffic signals, toll booths, and other infrastructure to help reduce congestion, prevent accidents, reduce emissions, and make transportation more efficient. Examples include fleet management, intelligent traffic management, vehicle-to-everything (V2X) communication, electric vehicle charging, electronic tolling, and a wide range of other mobility solutions.

Terminal operations: In a nonmechanized seaport, operator travel can take up to 40% of their working time, resulting in a significant drop in productivity, soaring costs, and deterioration in customer service. However, with the multiplication of intelligent solutions, the connection of machines to each other, and the development of information systems as well as the network of connected objects, it is very easy and often less expensive than one might think to automate all or part of its flows.

Today's port professionals need to have a thorough understanding of container terminal management, the developments that impact their development and the challenges and opportunities for container ports: growth in container traffic requiring additional container capacity, ever larger ships calling at your port and terminals, alliances between container lines, automation of port terminal operations, growing importance of intermodality, and increased competition focusing on profitability and service.

Smart container management: The automated container terminal is not a futuristic view. For over a decade, automation has been implemented in terminals, and this trend is continuing. Today, in some terminals, the management of container fleets (including storage and movement) is done with fully automated equipment. Automated vehicles such as automated guided vehicles (AGVs) and straddle carriers are used in some terminals for transporting containers between storage areas and dock gantries.

For terminals, IT management is an element of activity management (storage and movements) in relation to ship planning and inland service tools. The tools for inland services. With the total automation of the terminal, we can imagine totally autonomous terminal managing flows: positioning and provisioning. The "smart terminal" will become, over several hectares, what is done in some logistics warehouses. The future would then be a terminal with artificial intelligence for total management.

2.3 IoT model for smart maritime ports

The deployment of smart environment, such as smart city, smart home, etc., uses the approach of Internet-of-Things (IoT) models. In such models, a set of sensing nodes (or IoT end-devices) is dispatched in this environment to collect different types of measurements/data. These data are transmitted over different communication links and networks and then stored somewhere for further real time or offline treatment, exploration, and exploitation. The large success of the IoT leads engineers to adopt different IoT models, mostly a three-layer model, a four-layer model, or a five-layer model. Because this chapter focuses on vehicular applications in the port environment, we adopt the model presented in **Figure 3**, which is sometimes called "IoT value chain" and used in the literature to point out the services related to vehicular applications. This includes six levels:

- Sensing layer: This layer is constituted by different end-devices known as sensors/actuators, which sense the environment and collect data. These data can be either sent directly to the upper layer or be submitted to some local but limited processing, such as compression and/or filtering.

- Transport/communications layer: The data collected locally by the sensing layer are given to this layer to be transmitted, generally to far locations where we have some advanced data processing equipment. This layer can be built by different alternative technologies. This can be wired technology like optical fiber and cables, or wireless using IEEE technologies (WiFi, WiMAX, long-range Bluetooth, or ZigBee) or 3GPP transmission technologies, which are categorized in second generation (2G), third generation (3G), fourth generation (4G), or 5G.

- Data layer: The data collected by the sensing layer and transmitted arrive to this destination where they are stored for advanced processing. This layer is comprised of data centers or clouds where advanced big data operations are executed.

- Analytics layer: This layer is characterized by a higher level of intelligence, where algorithms or processes related to artificial intelligence and/or machine learning treat the stored data.

- Control layer: Based on the insights and knowledge gained from the analytics layer, this layer should meet optimal decisions to control the processes or the infrastructure monitored through the considered IoT system.

- Business layer: The IoT system is foreseen to be deployed in industrial and/or professional environments. Therefore, the objective remains generally the achievement of some economic benefits. The responsible processes to guide the entire processes to achieve these objectives and benefits must be englobed in this layer.

Figure 3.
Simple horizontal model for the IoT value chain in vehicular applications.

3. Vehicular applications/systems in smart maritime port environment

3.1 Vehicles in maritime port yard

The maritime ports is a very active area, where heavy goods mostly encapsulated in containers must be handled in different processes, namely, loading, unloading, storage, delivery toward hinterland, received from hinterland, controlling through customs, etc. These operations require the use of different vehicular machines such as cranes, mobile cranes on wheels or rails, cranes operated by drivers, automated cranes, trucks on port yard, hinterland trucks, trains from hinterland, etc. **Figure 4** shows some examples of vehicles on port yard, including unmanned aerial vehicles (UAV) or drones. Some ports also use underwater unmanned vehicles (underwater drones) for different control and environment supervising tasks or control the part ship container underwater.

Automated guided vehicles (AGV): An automated guided vehicle (AGV) or automatic guided vehicle is a vehicle used to transports containers (and other items) in the port's area. These vehicles are remotely controllable, but also have an advanced level of intelligence to execute self-driving and other optional decisions, as depicted in **Figure 5**. Several advantages are guaranteed through the use of AGVs, such as economic/cost effectiveness, time efficiency, increased efficiency of the overall processes in the ports, considerable reduction in accidents, and reduced injuries due to the absence of human traffic, etc.

- **UAV or drone-assisted operation:** Drone or unmanned aerial vehicles (also unmanned aircraft systems—UAS) are gaining a large application fields. Maritime ports are already testing drones for their operation, such as [4]:

- inspections of physical structures and patrolling of security rounds;

Figure 4.
Examples of vehicular applications in smart maritime port environments.

Figure 5.
Fleet of AGVs as part of port operations automation (Credits: wikiwand.com).

- routine inspections for maintenance of buoys, pipes, docks, breakwater cranes, roof ships, and other structures that are conventionally difficult to access;
- stock measurement to calculate (bulk) volumetric mass inventory. Inventory volume tracking of outdoor bulk material storage areas;
- detection of irregular situations, leaks, or abnormalities through (thermal and gas) sensors, as supplemental emergency support without the need to expose people to the affected areas;
- measurement and control of environmental aspects, detection of contamination, and tracking and monitoring those responsible for the environmental breaches. Environmental and ecological monitoring of both on-site processes and surrounding areas.

3.2 Levels of intelligence in vehicular applications in SMP

The different types of vehicles used in the port yard have the objective of facilitating the port operation through the optimization of processing time, safety of goods and personnel, and achieving cost and energy efficiency. However, higher levels of optimization are only reached through the automation of different operations, even also through the operation of vehicles in maritime ports. Therefore, we talk about autonomous or intelligent vehicles. The transition from traditional/classical vehicles towards autonomous port vehicles cannot be done in one transition. Thus, there are different levels of autonomy of vehicles, as illustrated in **Figure 6**. These intermediate levels allow a smooth transition, where the research and development works can be accompanied with the proof of concepts in the real life of maritime ports. This will allow the engineers get feedback from the practice to adapt their systems to any existing or new challenges in such a harsh and critical infrastructure. Higher autonomy signifies higher intelligence that requires the collection of larger amount of

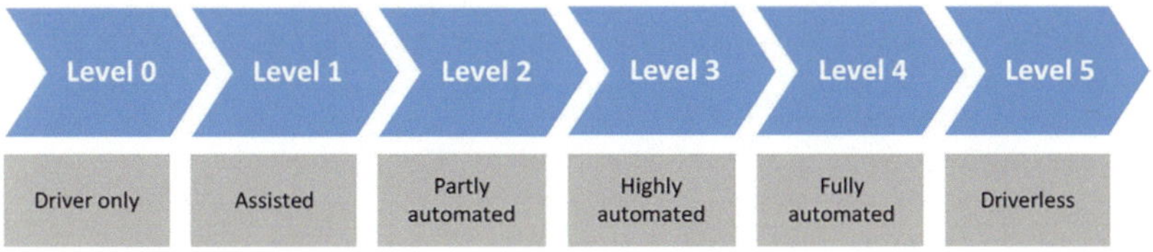

Figure 6.
Generic model for autonomy levels of vehicles in port operations [5].

data, which in its turn must be treated and/or transmitted in very short delay and in high reliable manner. In other words, these vehicles will need ultra-low latency and ultra-reliable communications links.

In a general context, the introduction of autonomous vehicles has three main objectives [5]:

- increased efficiency of transport, which brings alongside better capacity;
- utilization;
- less negative environmental impact;
- increased safety.

4. Vehicular communications in maritime ports

4.1 General hierarchical architecture/infrastructure in the smart port

From the economic point of view, it is worthy to select communication technology that can support all (or at least the maximum of) customer services. However, in practice, this is not possible. On the one hand, the diversity in services and the generated traffic have different quality-of-service requirements. On the other hand, each developed technology has its strengths and weaknesses, as can be seen in **Figure 7**, where only two performance criteria are plotted for different transmission technologies, namely, coverage and throughput. In reality, operators and engineers of the network must take into consideration further performance criteria such as delay, reliability, deployment as well as operational costs, etc.

An optimal design of the networking layer consists in having a networking infrastructure, which can support a variety of services from different vertical smart domains. A simplified topology of communication infrastructure in the area of a smart maritime port is illustrated in **Figure 8**. In such architecture, we recognize three levels, as discussed first in [6]:

- Level of low band or low bandwidth: This connects to the end-devices of the sensor layer (built through sensors and actuators). These nodes are dispatched in wide areas and transmit, in general, short and few parquets. Therefore, these nodes are not very requiring concerning the quality of services. The networking technologies to build this level must be very energy efficiency (low energy consumption) while offering low bit rates. Generally, wireless technologies such as Bluetooth, Bluetooth energy (BLE), ZigBee, Z-wave, low-power wide area

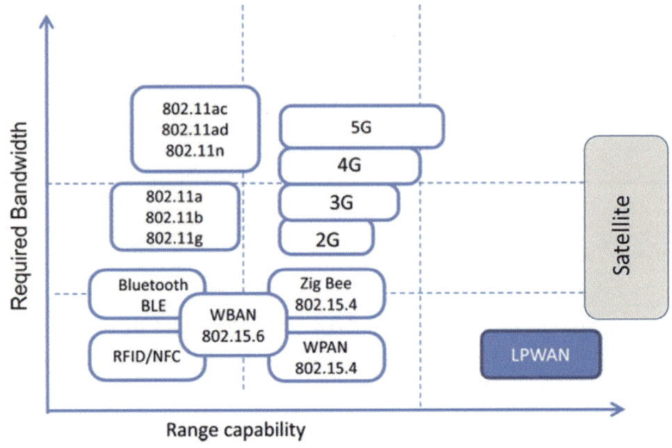

Figure 7.
Classification of wireless communications technologies in a two-dimensional plane (coverage and throughput).

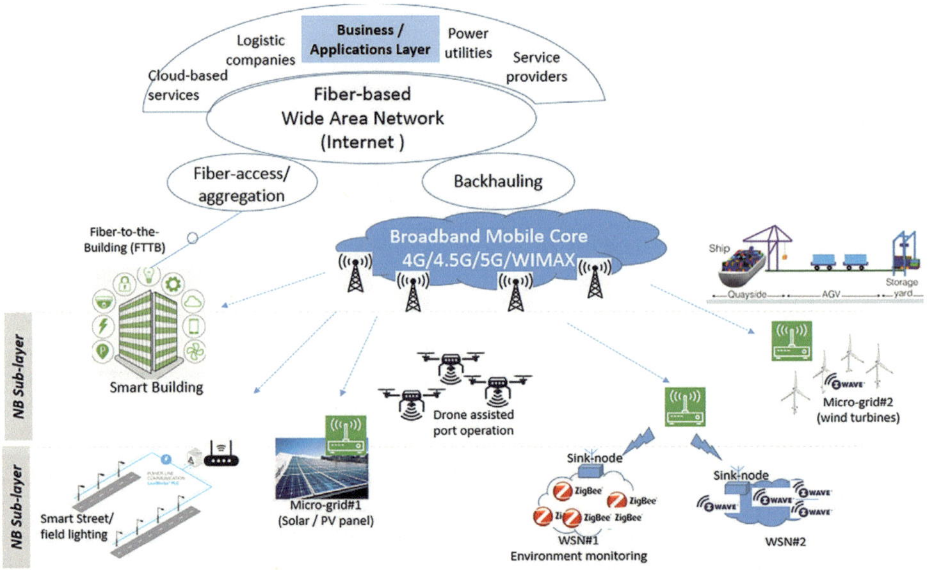

Figure 8.
Sublayers of communication architecture in a smart port.

(LPWA) networks (SigFox or LoRa or WightLess). As we remark, the wireless solutions are advantageous in this level, as they are easy to deploy and to maintain.

- Level of medium bandwidth: This level must transmit the traffic aggregated from the nodes of the lower level. Normally, for each node pool (or cluster of nodes), there is a gateway that collects and aggregates the traffic outgoing from these nodes. Because of that, the demand of the transmission capacity is relatively higher. The determination of the size of the clusters is subject to optimization under different constraints. But, we can consider that a bit rate of 2 Mbps a good compromise.

- Level of broadband technologies: In its turn, the level of medium bandwidth must be aggregated and connected to the (national) wide area networks. For most of the cases in the IoT, wireless solutions are more advantageous. Furthermore, mobility plays an important role here. Therefore, we have greater requirement for mobile broadband communications in this level. This is the mobile backbone of the networking layer in the smart port area. In addition, this level supplies direct connections to end-devices with higher demand of bandwidth, like HD camera, mobile PC/tables/device of the field force, normal Internet services everywhere in the port area, etc. The notion of mobile broadband has appeared in the engineering field related to third generation (3G) of mobile communications, technically known as universal mobile telecommunications system (UMTS). However, this was not able to offer transmission capacity more than 2 Mbps. Therefore, this has been extended to offer higher capacities and its derivation was known as high-speed downlink packet access (HSDPA) that focused on the improvement of downlink connection. This was followed by high-speed uplink packet access (HSUPA) that focused on the improvement of uplink connection. The last two variants have converged in a relative advanced technology called evolved high-speed packet access (eHSPA or HSPA+), which adopted advanced antenna technology as well as advanced modulation and coding schemes to reach up to 40 Mbps. The major (re)evolution in mobile broadband communications was realized by the fourth generation (4G) or long-term evolution (LTE), which can achieve bit rates over 300 Mbps. This milestone was rapidly overruled by the releases of long-term evolution-advanced pro (LTE-A Pro, called also 4.5G) reaching the magic threshold of 1 Gbps. This generation has offered different performance milestones in form of capacity of the bandwidth, shorter transmission delay, and service coverage for massive IoT systems. Nowadays, the fifth generation (5G) is struggling to find field where it can beat the performance of 4.5G. This seems to be a hard task, because different technical aspects as well as economic factors are putting lot of pressure on the deployment roadmaps of 5G. This has resulted in a delay in the widespread deployment of 5G technology.

4.2 Qualitative comparison of broadband technologies for vehicular applications

When we consider the evaluation of communication technologies for vehicular applications in smart port environment, a qualitative comparison is used, according to different criteria, which are throughput, network costs, coverage, adequacy to support massive IoT and the delay/latency. Here, the short-range communications technologies or short-range networks (SRNs) are excluded, because the main challenge of the network communications in port application lies in the set of (**large coverage, short delay, and sometimes broadband**). Furthermore, we have also taken into consideration the costs, which are very important for the network operator, the reliability as main challenge for sensitive application, and finally the IoT adaptability. This latter one is very important because any smart environment has a sensor or IoT layer as basis for any smart thing. Therefore, the communication layer should be adequately chosen for such environment.

Long-term evolution/advanced (LTE/LTE-A) is the first real mobile broadband technology. Having a real success worldwide, this is the backbone for the current mobile network and mobile Internet. Its major strength is its characteristic of "long-term" evolution; thus, it was designed to be able to be extended anytime for any new services. Indeed, the first LTE releases (Release 8 and Release 9) have been quickly evolved to Release 10 and Release 11, etc. Major transitions have realized in a short time, so LTE overrides just usual broadband Internet to support machine-type communication

(MTC), machine-to-machine (M2M), vehicle-to-vehicle (V2V), Internet-of-Things (IoT) communication/services/applications. This lets this technology show higher performance in all criteria compared with other technologies. Of course, it has a unique weakness, which is the coverage compared to the satellite. A very large number of research works have investigated the performance of the LTE in vehicular communications either in the form of V2V, vehicle-to-infrastructure (V2I), vehicle-to-everything (V2X), including Pedestrian, or Internet of Vehicles (IoV) [7–11]. Typical types of utilization of vehicular communications are depicted in (**Figure 9**) for general use as well as for use in the case of maritime port environment by mobile work force.

5G is the newest mobile technology generation. It should bring higher performance than 4G, especially in three dimensions: enhanced broadband throughput, ultrareliable communication, and support of massive IoT devices. The enhanced broadband should support new applications requiring very high capacities, such as virtual reality/augmented reality (VR/AR), which are not expected to be required in smart port environments. A major weakness is relative high cost to be invested. This is because this generation requires the deployment of new infrastructure. In the opposite of the evolved releases of 4G, which the upgrade of the networks is made with less costs. Different works evaluated the use of 5G for vehicular applications, and some of them are even talking about 6G [9, 14–16]. An example of application of 5G for vehicular applications in maritime port environment is given in [17], where higher capacity and lower latency are the objectives of the adopted architecture (**Figure 10**). The competition between 4G/LTE and 5G to build network layer for smart maritime ports will be intensive, because they are on the same level of performance in most of the considered criteria, as illustrated in **Figure 11**. However, 4G has the main advantage of network costs, because 4G network are already massively deployed and proofed in the field, while 5G massive rollout suffer from delay, viability of its business model and other different economic circumstances.

The satellite communication is the excellent solution for the large coverage (everywhere and anytime). This technology did not cease to evolve over the years. However, two major weak points are inevitable: the high costs for broadband communications and the long delay, which is caused by the signal propagation to the sky and back. In spite of this, different mechanisms have been investigated to improve its performance [18–21].

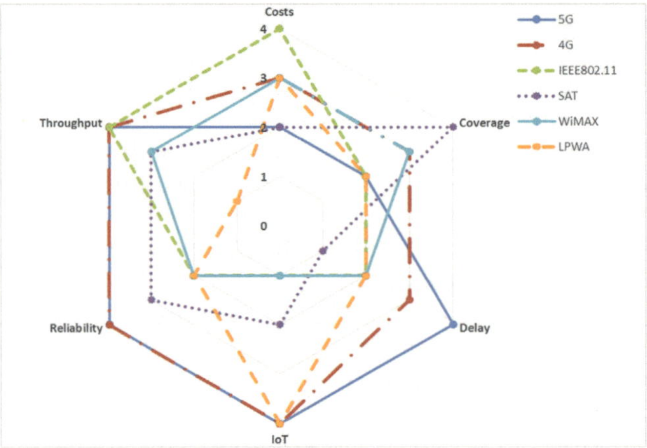

Figure 9.
Types of V2X in LTE standards (left, [12]) and its applications in maritime port environment (right, [13]).

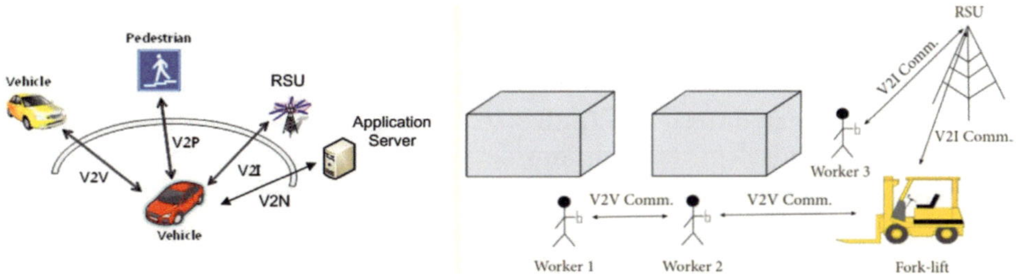

Figure 10.
Example from the practice for the use of 5G in maritime port vehicular applications [17].

To realize the objective of long range or long coverage of IoT applications and devices, without the use of satellite, low-power wide area (LPWA) networks have been developed especially for the narrowband IoT. Thus, high throughput and implicitly the delay are the main drawbacks of this technology. Examples of the utilization of LPWA in vehicular applications can be found in [22–25].

WiMAX technology has been reserved for some industrial applications, after it lose the battle of 4G, and this is the adoption of the usual broadband network for the public mobile network. This worked well for a while, however, with the apparition of massive IoT applications in industrial fields under the name of Industrie 4.0; WiMAX was not able to support the modern services, because its standardization has been frozen more than a decade ago. Therefore, it cannot be a good favorite or alternative for such modern smart environment like smart maritime ports.

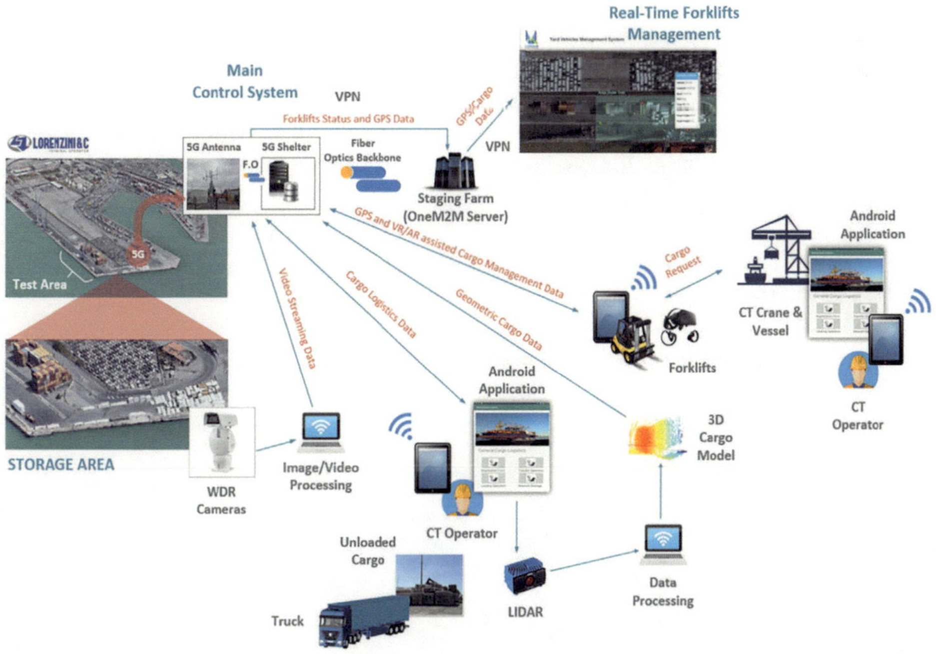

Figure 11.
Qualitative performance of communications technology (relative to each other) for vehicular applications (scores meaning: 4—very good, 3—good, 2—average/acceptable, 1—bad, and 0—very bad).

The success of the WLAN standard (IEEE 802.11) has led to its extension beyond residential and enterprise areas. Special applications have been developed, including for smart cities (IEEE 802.11h) and smart transportation (IEEE 802.11p). While studies have investigated the use of IEEE 802.11p for vehicular applications, such as [26–30], there is a lack of research considering its use in the specific context of a smart maritime port environment.

5. Conclusions

The maritime transport networks are playing a unique and major role in an always more globalized world economy. In such networks, the maritime ports play the role of hubs. Any disturbances in these hubs will negatively influence the worldwide economy Therefore, economy players are transforming the ports through an evolving process to become smart maritime ports. These smart ports are built through an ensemble of smart domains that are adopting sensing, data transmission, and data intelligence to support intelligent decision-making processes. Smart grid/microgrids, smart container management, smart/automated terminal operations, etc. are examples of such smart domains.

In different smart domains, vehicular applications are present. Therefore, the vehicular communications must be taken into consideration while developing the roadmap for smart maritime port. In this chapter, we focused on five communication technologies through the presentation of a qualitative performance evaluation of these technologies. The introduction of autonomous imposes hard requirements for the quality of services to be guaranteed by the communication networks, such as ultra-low latency, ultra-high reliable communication links.

Acknowledgements

The authors would like to thank our colleagues in the department for their lecture, corrections and valuable feedbacks that helped to improve the scientific quality of this chapter.

Conflict of interest

The authors declare no conflict of interest.

Author details

Abdelfatteh Haidine[1], Ayoub El Idrissi[1], Abdelmoula Ait-Allal[2*], Aziz Dahbi[1] and Abdelhak Aqqal[1]

1 National School of Applied Sciences, University Chouaib Doukkali, El Jadida, Morocco

2 High Institute of Maritime Studies (ISEM), Casablanca, Morocco

*Address all correspondence to: haidine.a@ucd.ac.ma

IntechOpen

References

[1] Bodewig K. Shaping the Future Policy of the European Maritime Space Motorways of the Sea Detailed Implementation Plan of the European Coordinator: Report EU Mobility and Transport. Brussels, Belgium: European Commission; 2020

[2] Miller G. Record shattered: 73 container ships stuck waiting off California, FREIGHTS WAVES News on Thursday, September 16, 2021. Available from: https://www.freightwaves.com/news/record-shattered-61-container-ships-stuck-waiting-off-california

[3] Spangler W, Kreulen J, Chen Y, Proctor L, Alba A, Lelescu A, et al. A smarter process for sensing the information space. IBM Journal of Research and Development. 2010;**54**: 1-13. DOI: 10.1147/JRD.2010.2050541

[4] Bexiga V. Blue Innovation—Drones in Port Operations, Escola Europa Intermodal Transport. 2021. Available from: https://escolaeuropea.eu/odiseo/issue-30-spring-2019/blue-innovation-drones-in-port-operations [Accessed: June 23, 2021]

[5] Fiedler R, Bosse C, Gehlken D, Brümmerstedt K, Burmeister H-C. Autonomous Vehicles' Impact on Port Infrastructure Requirements. Hamburg: Fraunhofer Center for Maritime Logistics and Services CML; 2019

[6] Haidine A, Ait-Allal A, Aqqal A, Dahbi A. Networking layered for the evolution of maritime ports into a smart environment. In: The International Archives of Photogrammetry, Remote Sensing and Spatial Information Sciences; Gottingen. Gottingen: Copernicus GmbH; 2021. pp. 251-257. DOI: 10.5194/isprs-archives-XLVI-4-W5-2021-251-2021

[7] Bazzi A, Cecchini G, Zanella A, Masini BM. Study of the impact of PHY and MAC parameters in 3GPP C-V2V mode 4. IEEE Access. 2018;**6**:71685-71698. DOI: 10.1109/ACCESS.2018.2883401

[8] Izydorczyk T, Tavares FML, Berardinelli G, Bucur M, Mogensen P. Performance evaluation of multi-antenna receivers for vehicular communications in live LTE networks. In: 2019 IEEE 89th Vehicular Technology Conference (VTC2019-Spring). Kuala Lumpur, Malaysia. 2019. pp. 1-6. DOI: 10.1109/VTCSpring.2019.8746609

[9] Muhammad M, Safdar GA. Survey on existing authentication issues for cellular-assisted V2X communication. Vehicular Communications. 2018;**12**:50-65. DOI: 10.1016/j.vehcom.2018.01.008. Available from: https://www.sciencedirect.com/science/article/pii/S2214209617302267

[10] Sliwa B, Patchou M, Wietfeld C. Lightweight simulation of hybrid aerial- and ground-based vehicular communication networks. In: 2019 IEEE 90th Vehicular Technology Conference (VTC2019-Fall). Honolulu, Hawaii. 2019. pp. 1-7. DOI: 10.1109/VTCFall.2019.8891340

[11] Gyawali S, Xu S, Qian Y, Hu R. Challenges and solutions for cellular based V2X communications. IEEE Communications Surveys & Tutorials. 2020;**2020**:1-1. DOI: 10.1109/COMST.2020.3029723

[12] European Telecommunications Standardization Institute (ETSI). LTE, Service Requirements for V2X Services. ETSI Technical Specification for (3GPP TS 22.185 version 14.3.0 Release 14).

Available from: http://www.etsi.org/standards-search

[13] Sanja B, Jose Angel L, Schmeink A, Azam S, Mathar R. On deploying vehicular communications at the developing seaport and related innovation success impediment. Journal Transport. 2019;**34**:126-134. DOI: 10.3846/transport.2019.7987

[14] Molina-Masegosa R, Gozalvez J. LTE-V for sidelink 5G V2X vehicular communications: A new 5G technology for short-range vehicle-to-everything communications. IEEE Vehicular Technology Magazine. 2017;**12**(4):30-39. DOI: 10.1109/MVT.2017.2752798

[15] Zeadally S, Javed MA, Hamida EB. Vehicular communications for ITS: Standardization and challenges. IEEE Communications Standards Magazine. 2020;**4**(1):11-17. DOI: 10.1109/MCOMSTD.001.1900044

[16] Cheng X, Huang Z, Chen S. Vehicular communication channel measurement, modelling, and application for beyond 5G and 6G. IET Communications. 2020;**14**:3303-3311. DOI: 10.1049/iet-com.2020.0531/*Fgiure port

[17] Tesei A, Lattuca D, Tardo A, Di Mauro L, Pagano P, Luise M, et al. Securing seaport logistic vehicles using a distributed ledger-based credential management system. IEEE Open Journal of Vehicular Technology. 2021;**2021**:2. DOI: 10.1109/OJVT.2021.3067209

[18] Cui G, Long Y, Xu L, Wang W. Joint offloading and resource allocation for satellite assisted vehicle-to-vehicle communication. IEEE Systems Journal. 2020;**2020**:1-12. DOI: 10.1109/JSYST.2020.3017710

[19] Yin Z, Jia M, Cheng N, Wang W, Lyu F, Guo Q, et al. UAV-assisted physical layer security in multi-beam satellite-enabled vehicle communications. IEEE Transactions on Intelligent Transportation Systems. 2021;**23**(3):2739-2751

[20] Lei A, Ogah C, Asuquo P, Cruickshank H, Sun Z. A secure key management scheme for heterogeneous secure vehicular communication systems. ZTE Communications. 2019;**14**(S0):21-31

[21] Yuhan R, Rui Z, Yongzhao L, et al. Spectral-energy efficiency tradeoff in cognitive satellite-vehicular networks towards beyond 5G. In: 2019 IEEE Wireless Communications and Networking Conference (WCNC). 2019. pp. 1-6

[22] Yuke L, Shuangshuang H, Linyao Y, et al. LoRa on the move: Performance evaluation of LoRa in V2X communications. In: IEEE Intelligent Vehicles Symposium (IV). Suzhou, China. 2018. pp. 1107-1111

[23] Akira S, Yoshitaka S, Toshihiro T. Evaluation of performance on lpwa network realizes for multi-wavelength cognitive v2x wireless system. In: 2019 IEEE 10th International Conference on Awareness Science and Technology (iCAST). Morioka, Japan: IEEE; 2019. pp. 1-6

[24] Jinhyung O, Dong-Woo L, Kyu-Min K. Unmanned aerial vehicle identification success probability with LoRa communication approach. In: IEEE 31st Annual International Symposium on Personal, Indoor and Mobile Radio Communications. London, United Kingdom. 2020. pp. 1-6

[25] Fei G, Jianwei N, Landu J, et al. Survey of the low power wide area network technologies. Journal of Network and Computer Applications. 2020;**149**:102459

[26] Singh A, Singh B. A study of the IEEE802.11p (WAVE) and LTE-V2V technologies for vehicular communication. In: International Conference on Computation, Automation and Knowledge Management (ICCAKM). 2020. pp. 157-160. DOI: 10.1109/ICCAKM46823.2020.9051468

[27] Singh PK, Nandi SK, Nandi S. A tutorial survey on vehicular communication state of the art, and future research directions. Vehicular Communications. 2019;**18**:2214-2096. DOI: 10.1016/j.vehcom.2019.100164

[28] Trujillo-Flores JI, Torrealba-Meléndez R, Muñoz-Pacheco JM, Vásquez-Agustín MA, Tamariz-Flores EI, Colín-Beltrán E, et al. CPW-fed transparent antenna for vehicle communications. Applied Sciences. 2020;**10**:6001. DOI: 10.3390/app10176001

[29] Arena F, Pau G, Severino A. A review on IEEE 802.11p for intelligent transportation systems. Journal of Sensor and Actuator Networks. 2020;**9**:22. DOI: 10.3390/jsan9020022

[30] Klapez M, Grazia CA, Casoni M. Minimization of IEEE 802.11p packet collision interference through transmission time shifting. Journal of Sensor and Actuator Networks. 2020;**9**:17. DOI: 10.3390/jsan9020017

Section 2

Advanced Approaches

Chapter 4

Secure Vehicular Networking: Architectures, Applications, Attacks, and Challenges

Mariyam Ouaissa, Mariya Ouaissa and Zakaria Boulouard

Abstract

The rapid advancement in technology has significantly impacted the field of communication, particularly with the evolution of wireless technology. Among these developments, Vehicle-to-Everything (V2X) communications have emerged as a pivotal technology for intelligent transportation systems (ITS). This innovation facilitates a network wherein vehicles communicate with each other and their surroundings, sharing vital information about traffic conditions and potential hazards. This network is instrumental in enhancing traffic management through route optimization and congestion control, while also enriching passenger experience with on-board infotainment options. Designed with a focus on road safety and passenger comfort, V2X networks are equipped with advanced on-board sensors capable of detecting various scenarios such as road accidents, traffic congestions, and other obstacles. This chapter delves into the intricate details of vehicular communications, exploring their architectures, characteristics, applications, and governing standards. Furthermore, it elucidates the concept of communication security within these networks, outlining the existing models of attackers, classifications of attacks, and fundamental security mechanisms. The comprehensive overview aims to shed light on the critical aspects of securing vehicular networks in the ever-evolving landscape of communication technology.

Keywords: ITS, V2X, RSU, IDS, security services, attacks

1. Introduction

In response to evolving transportation demands, traditional transport management technologies are proving to be less effective. Numerous studies have highlighted the importance of integrating intelligent transportation systems (ITS) into infrastructure to meet future transportation needs cost-effectively [1]. ITS, utilizing advanced wireless communications and localization systems, are set to pave the way for innovative solutions that foster interactions among drivers, vehicles, and the road environment [2]. This concept is further enhanced by the advent of autonomous on-board devices augmented with wireless communications, leading to the rise of the "connected vehicle" paradigm. This new approach in intelligent transport systems focuses on enhancing road traffic safety and efficiency through wireless communications.

Connected vehicle communications, known as V2X, includes wireless interactions such as Vehicle-to-Infrastructure (V2I), Vehicle-to-Vehicle (V2V), and hybrid communication forms [3].

Being distributed systems, vehicular networks encounter numerous security and privacy challenges, primarily because the data transmitted are exposed in an open-access environment. In particular, to fulfill the core objectives of V2X, especially in road safety applications, it is imperative to have robust security measures in place to facilitate the smooth functioning of V2X systems [4]. Absent these measures, V2X systems risk is being exploited for compromising security and disrupting traffic management [5]. Consequently, there is an urgent need for more effective strategies to safeguard the security and privacy of these networks against potential attacks, with a focus on protecting the confidential information of drivers and passengers [6].

This chapter provides an in-depth analysis of contemporary advancements in both research and the establishment of standards pertaining to the security and privacy elements in Vehicle-to-Everything (V2X) communication. The goal is to offer a comprehensive synopsis of vehicular communication systems, highlighting their structural designs, distinctive features, various applications, and associated standards. Subsequently, the discussion shifts to the principles of secure communication as implemented in vehicular networking. To conclude, an examination of security within these networks is undertaken, encompassing the exploration of potential attacker profiles, categorization of different types of attacks, and an introduction to fundamental security protocols.

This chapter is structured as following: Section 2 provides a broad overview, focusing on the primary components and architectures integral to vehicular communications. Following this, the distinct features of these systems are elaborated in Section 3, while Section 4 is dedicated to discussing their various applications. The standards governing vehicular communications are then detailed in Section 5. Moving on to Section 6, the paper addresses the various challenges associated with vehicular communications. Section 7 delves into the realm of network security, discussing the existing models for potential attackers, classifications of different types of attacks, and the fundamental security mechanisms in place. The paper concludes with a final section that synthesizes the key points and findings.

2. Components and architectures of vehicular networks

The implementation of vehicular networks is essential, particularly with the integration of V2X technology. This technology is notable for its adaptability to fluctuating network topologies, allowing nodes to seamlessly join or leave the network without compromising its overall functionality or diminishing its performance [7]. These networks are characterized by their autonomous and self-organizing attributes, functioning as wireless communication systems where vehicles serve as mobile nodes. In this setup, vehicles are equipped with On-Board Units (OBUs), facilitating not only inter-vehicle communication but also interactions between the vehicles and the Road Side Units (RSUs). This dynamic communication framework is instrumental in enhancing the efficiency and efficacy of modern transportation systems [8].

2.1 Communicating entities

2.1.1 On-Board Units

An On-Board Unit is a wave device typically installed within a vehicle. It serves the primary function of facilitating information exchange between different OBUs or with Road Side Units. Comprising a combination of advanced hardware and software components, such as Global Positioning System (GPS), radar, cameras, and various sensors, the OBU represents a hub of high-tech integration in vehicular communication systems. Central to its design is a transceiver, which typically includes a radio frequency antenna connected to a processor. This sophisticated arrangement enables the OBU to perform its critical role in vehicular networks, ensuring efficient and effective communication across different nodes within the system.

2.1.2 Road Side Unit

Road Side Units, similar to OBUs, are composed of an antenna, a processor, and read/write memory. These units are equipped with both wired and wireless interfaces to facilitate diverse communication needs. The wireless interface is primarily used for communication with vehicle OBUs, enabling direct interactions between the vehicle and the roadside infrastructure. On the other hand, the wired interface is utilized to establish connections with other RSUs and the Internet, thus ensuring a broader range of connectivity.

Additionally, RSUs play a crucial role in expanding the coverage area of OBUs by transferring data, effectively enhancing the communication network's reach and reliability. Strategically installed along roadways, RSUs are commonly found in key locations such as near intersections and gas stations. This placement is instrumental in maximizing their effectiveness in transmitting and receiving vital information, thereby significantly contributing to the overall efficiency and safety of vehicular networks.

2.1.3 The central authority

The central authority is the pivotal entity in managing vehicular networks, serving as the network's backbone and acting as a data storage server. Its primary responsibilities encompass the comprehensive management of the entire network, which includes the registration of RSUs, OBUs, and vehicle users. This role is crucial for maintaining an organized and efficient network.

In addition to these administrative duties, the central authority is also tasked with ensuring the secure operation of Vehicle-to-Everything communication. This involves verifying vehicle authentication, as well as the identification of both users and OBUs. Such verification processes are essential for preventing any potential harm or damage to vehicles within the network.

Moreover, the central authority is equipped with specialized mechanisms designed to identify potential attackers. This capability is vital for safeguarding the network against various security threats, thereby maintaining the integrity and reliability of vehicular communications. By fulfilling these roles, the central authority plays a key role in ensuring the smooth and secure functioning of the vehicular network ecosystem.

2.2 Communication architectures in V2X

The primary objective of a Vehicle-to-Everything architecture is to facilitate seamless communication within the vehicular network [9]. This includes enabling interactions not only between vehicles but also between vehicles and stationary road infrastructure [10]. The communication framework in V2X encompasses three key modalities:

2.2.1 Vehicle-to-Vehicle communication

Vehicle-to-Vehicle communication is centered around the wireless exchange of data directly between vehicles. The primary aim of V2V communication is to enhance road safety by enabling vehicles in transit to share critical information such as their position, speed, emergency braking actions, potential collisions, and deceleration instances. This direct form of communication allows for real-time data exchange, which is crucial for preventing accidents and improving the overall safety of vehicular traffic.

One of the key advantages of V2V communication is its efficiency in multi-hop ad-hoc communication, which often proves to be more effective than relying on a network facilitated by operators. In multi-hop ad-hoc communication, data can be transmitted from one vehicle to another in a relay-like manner, allowing for a wider range of communication without the need for a fixed infrastructure.

Figure 1 illustrates this mode of communication, demonstrating how vehicles can communicate with each other directly. This depiction would show the flow of information between vehicles, highlighting the dynamic and decentralized nature of V2V communication. Such a system is instrumental in creating a more responsive and interconnected vehicular network, where each vehicle acts both as a transmitter and a receiver, collaboratively contributing to a safer driving environment.

2.2.2 Vehicle-to-Infrastructure communication

Vehicle-to-Infrastructure communication refers to the method where multiple vehicles communicate with a range of roadside devices that are integral to a country's road network. This mode of communication involves various types of infrastructure units, such as radio frequency identification (RFID) readers, traffic signs, cameras, traffic lights, street lights, and parking meters. These devices play a critical role in providing

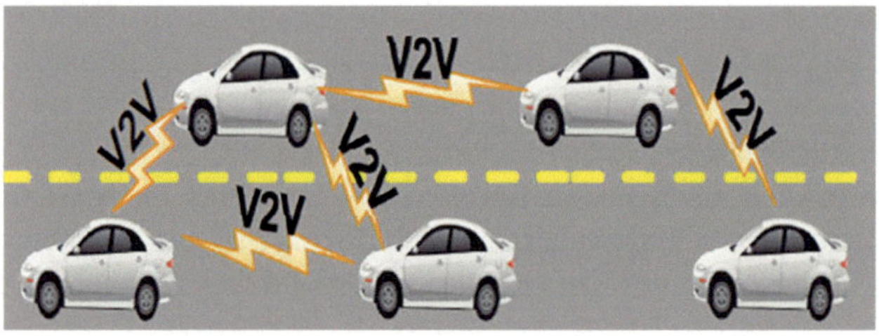

Figure 1.
Vehicle-to-Vehicle (V2V) communication.

vehicles with essential information about the road environment, traffic conditions, and other relevant aspects.

The V2I communication technique is predominantly wireless and bidirectional. This means that the data transfer occurs in both directions: infrastructure units can easily transmit information to vehicles via a wireless network, and vehicles can similarly send information back to these units. This two-way communication allows for a dynamic exchange of information, enhancing the overall efficiency and safety of the transportation system.

Figure 2 demonstrates how this communication between vehicles and infrastructure takes place. It would typically illustrate the flow of information from various roadside devices to the vehicles and vice versa, showcasing the interconnected nature of V2I communication. This visualization helps in understanding how V2I technology facilitates a smarter, more responsive road network, where the infrastructure and vehicles work in synergy to optimize traffic flow, reduce congestion, and improve road safety.

2.2.3 Hybrid communication

Hybrid communication in the context of vehicular networks is a comprehensive approach that integrates the capabilities of both Vehicle-to-Vehicle and Vehicle-to-Infrastructure communication. This combination allows for a more versatile and far-reaching communication network within the vehicular ecosystem.

In a hybrid communication setup, a vehicle has the ability to interact with road infrastructure using either single-hop or multi-hop modes. The single-hop mode involves direct communication between the vehicle and the infrastructure, while the multi-hop mode allows for the transmission of data across multiple vehicles or infrastructure points, effectively extending the communication range. This versatility enables vehicles to maintain long-distance connections, not only with the Internet but also with other vehicles that might be located at significant distances.

Figure 3 illustrates this hybrid communication model. It would typically depict how vehicles can seamlessly switch between or combine V2V and V2I communication modes, thereby facilitating a more connected and efficient network. This figure likely shows the flow of information between vehicles and infrastructure units, and possibly between vehicles themselves, highlighting the interconnected nature of hybrid communication. Such a system enhances the potential for improved traffic management, road safety, and overall user experience by leveraging the strengths of both V2V and V2I communication.

Figure 2.
Vehicle-to-Infrastructure (V2I) communication.

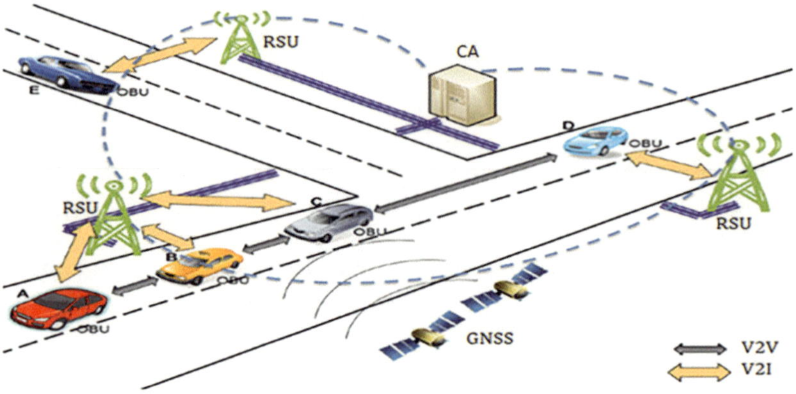

Figure 3.
Hybrid communication.

3. Characteristics of V2X

In the following, we will delve into a range of characteristics intrinsic to vehicular networks that are pivotal to consider when devising new solutions. These considerations are crucial for ensuring compliance with and effectively meeting the requirements of Vehicle-to-Everything standards. These characteristics encompass various technical, operational, and regulatory aspects that are fundamental to the successful implementation and functionality of V2X systems. By thoroughly understanding and addressing these characteristics, proposed solutions can be more accurately tailored to fit within the framework of V2X standards, thereby enhancing the efficiency, safety, and overall effectiveness of vehicular networks [11].

3.1 High mobility and dynamic topology

Vehicle-to-Everything systems are distinguished by their dynamic nature, primarily due to the vehicles constantly changing their positions and speeds. This rapid and frequent mobility results in a highly dynamic network topology, presenting unique challenges for maintaining stable and effective communication.

The swift movement of vehicles not only affects the structure of the network but also has a significant impact on wireless signal quality. High speeds can lead to fluctuations in signal strength, potentially resulting in packet loss or increased communication delays. These challenges are further compounded by varying traffic conditions. In urban areas, for instance, the density of vehicles can lead to congested networks, while on highways or in rural areas, the dispersion of vehicles creates a different set of communication dynamics.

Such variability in vehicle density and mobility patterns necessitates robust and adaptive communication systems within V2X networks. Solutions must be capable of handling the rapid changes in network topology and signal quality to ensure consistent and reliable communication, essential for the safety and efficiency of vehicular networks.

3.2 Availability of geographic location

In Vehicle-to-Everything systems, many geographic protocols operate on the premise that each vehicle is equipped with a positioning system integrated into digital maps. This integration is crucial for the effective functioning of these protocols, as it allows each vehicle to be aware of its precise location at any given time.

The Global Positioning System (GPS) is the most widely utilized positioning technology in vehicular networks. Its popularity stems from its ability to provide real-time, three-dimensional positional data, including latitude, longitude, and altitude. Additionally, GPS systems are capable of delivering accurate information about the vehicle's direction and speed.

This precise location and movement data are fundamental to the V2X communication framework. They enable vehicles to make informed decisions based on their immediate environment, contributing to safer and more efficient navigation. The integration of GPS with digital mapping technologies thus forms a cornerstone of modern vehicular network systems, underpinning various safety and navigation applications within the V2X ecosystem.

3.3 Mobility models

In vehicular network systems, the mobility model plays a crucial role in assessing the behavior and performance of various protocols. This model is pivotal as it simulates real-world vehicular movement patterns and conditions, allowing for a more accurate evaluation of network protocols within a V2X context.

To effectively reflect real-world conditions, the mobility model must account for various factors that influence vehicular movement. These include variations in speed, the presence of traffic lights, intersections, and the occurrence of traffic jams. Incorporating these elements helps in creating a more realistic simulation environment, which is essential for accurately predicting how protocols will perform in actual vehicular network scenarios.

The process of defining an appropriate mobility model begins with a clear understanding and identification of the test scenario environment. This involves considering the specific characteristics of the environment where the vehicular network will operate, such as urban settings with dense traffic and frequent stops or highway scenarios with higher speeds and fewer interruptions. By tailoring the mobility model to these specific conditions, it becomes possible to more accurately assess the efficacy and reliability of various V2X communication protocols under realistic conditions. This approach ensures that the protocols developed are robust and capable of handling the complex dynamics of real-world vehicular networks.

3.4 Energy and storage capacity

In Vehicle-to-Everything networks, the nodes—which include both On-Board Units in vehicles and Road Side Units—are well-equipped in terms of energy and computing power. This abundance of resources allows these nodes to effectively handle storage and processing tasks without significant limitations.

A key factor contributing to this capability is the lack of constraints on energy consumption for V2X nodes. OBUs, installed in vehicles, draw their power directly from the vehicle's own power supply, ensuring a steady and reliable energy source. This direct power supply allows OBUs to perform their functions without being hindered by energy conservation concerns typically faced by battery-powered devices.

Similarly, RSUs, which are installed as part of road infrastructure, are powered directly by the road platforms. This setup provides a stable and continuous energy source, allowing RSUs to operate efficiently and maintain consistent communication with passing vehicles.

The robust computing power and unfettered energy access of these V2X nodes ensure that they can effectively manage the demands of processing, data storage, and communication, which are critical for the smooth functioning of vehicular networks. This lack of energy constraints is a significant advantage, as it allows for more complex computations and longer operation times, leading to more reliable and efficient vehicular communication systems.

3.5 Quality of Service requirements

In Vehicle-to-Everything systems, the Quality of Service (QoS) requirements are highly diverse and depend greatly on the specific type of service being provided. Different applications within V2X demand varying levels of QoS to function effectively, particularly when it comes to real-time applications.

Real-time applications, especially those related to road safety and traffic management, have particularly stringent QoS requirements. These services necessitate guaranteed access to communication channels to ensure timely and reliable transmission of critical information. The nature of these applications means that any delay or disruption in communication could have serious consequences, making reliability and responsiveness top priorities.

Key among these QoS parameters are the requirements for end-to-end delay and packet loss rate. End-to-end delay refers to the time taken for a packet of data to travel from the source to the destination. For road safety applications, this delay needs to be minimized to ensure that vehicles can respond to dynamic road conditions in real-time. Similarly, the packet loss rate—the rate at which data packets are lost during transmission—must be kept as low as possible. High packet loss rates can lead to incomplete or inaccurate information being received, which is unacceptable in scenarios where safety-critical decisions are being made.

Therefore, in V2X communication, ensuring that these QoS requirements are met is crucial for the successful operation of real-time applications related to road safety and traffic management. This involves sophisticated network management and robust communication protocols that can prioritize and handle such critical data efficiently and reliably.

4. Applications of vehicular communications

V2X networks have been primarily developed with the goal of enhancing traffic safety and efficiency, primarily by minimizing the risk of road accidents. This overarching objective leads to the classification of V2X applications into two key categories based on their primary focus: security and efficiency. However, it's important to recognize that safety and effectiveness in V2X applications are not mutually exclusive concepts; instead, they are interconnected and often influence each other.

The aspect of security in V2X applications encompasses measures and protocols aimed at preventing accidents, ensuring the integrity and confidentiality of data, and protecting the network from various cyber threats. This safety aspect is crucial, as it directly impacts the well-being of the vehicle occupants and other road users.

On the other hand, efficiency in V2X applications refers to optimizing traffic flow, reducing congestion, and enhancing the overall driving experience. This includes intelligent traffic management, route optimization, and real-time traffic information dissemination, all of which contribute to smoother and more efficient road travel.

While these two categories can serve as a basis for classifying V2X applications, it's essential to approach their design by considering both safety and efficiency together, along with other factors like user experience, scalability, and environmental impact. Integrating these considerations ensures that V2X applications not only address immediate safety concerns but also contribute to a more sustainable, efficient, and user-friendly transportation ecosystem. This holistic approach in the design and development of V2X applications is crucial for maximizing their benefits and ensuring their long-term success in improving road safety and transportation efficiency [12].

4.1 Security applications

The primary objective of safety applications within Vehicle-to-Everything networks is to prevent and decrease the occurrence of road accidents. These applications fall under the category of delay-sensitive applications, where minimizing communication delay is crucial for effective operation.

To achieve reduced delay, safety applications in V2X predominantly utilize V2V communication. This direct form of communication between vehicles is faster and more efficient in disseminating critical information in real-time. In the event of an accident, the involved vehicle can instantly broadcast information about the hazardous situation to nearby vehicles. This immediate transmission of data allows approaching vehicles to become aware of the danger ahead in time to take necessary precautions or alter their routes.

Another essential requirement for these safety applications is reliability. It is vital that all vehicles in proximity to the potential danger are reliably informed of the situation. This ensures that every driver in the vicinity has the opportunity to react appropriately, thereby significantly enhancing road safety. The effectiveness of these safety applications hinges on their ability to consistently deliver accurate and timely warnings to all relevant parties, underscoring the importance of both speed and reliability in V2X safety systems.

4.2 Traffic management applications

V2X services extend beyond the realm of road safety, encompassing a broad spectrum of applications, particularly in the area of traffic management. These applications are designed with a focus on enhancing the efficiency of traffic flow and optimizing travel routes. By doing so, they contribute significantly to reducing the time vehicles spend on the road, which in turn leads to decreased fuel consumption and a reduction in air pollutants.

Traffic management applications within V2X systems utilize real-time data and advanced communication technologies to analyze traffic patterns, identify congestion points, and suggest alternative routes. This proactive management of traffic helps in alleviating congestion on busy roads and highways, leading to a smoother flow of vehicles. The ripple effects of these improvements are manifold, including a more pleasant driving experience, reduced wear and tear on vehicles, and lower emissions due to less idling and stop-and-go traffic.

An additional, yet crucial benefit of these traffic management applications is the potential reduction in road accidents. By effectively managing traffic flow and reducing congestion, the likelihood of accidents caused by traffic jams and the associated frustration or aggressive driving behaviors is significantly diminished. This not only contributes to overall road safety but also aids in creating a more sustainable and environmentally friendly transportation ecosystem.

Thus, V2X services, through their diverse range of applications, play an integral role in shaping a more efficient, safer, and cleaner transportation system, extending their impact well beyond just road safety.

4.3 Comfort or entertainment applications

The primary aim of many V2X applications is to enhance the comfort and convenience experienced by drivers and passengers. These applications offer a variety of services that cater to the informational and entertainment needs of those on the road. Examples include providing route information, locating available parking spots, ensuring internet connectivity, and offering a range of multimedia services such as messaging, gaming, and access to various radio channels.

One of the key aspects of these applications is the provision of Internet access, which is typically achieved through Vehicle-to-Infrastructure communications. This connectivity allows vehicles to tap into a wide array of online services, essentially bringing the full spectrum of business services into the vehicle environment. Such access transforms the vehicle into a mobile office or entertainment hub, enabling passengers and drivers to remain productive or entertained throughout their journey.

Furthermore, file sharing and video streaming services can be facilitated via Vehicle-to-Vehicle communications. This feature is particularly beneficial during long journeys, as it allows passengers to stream videos or share files with other vehicles, thus making travel more enjoyable and less monotonous.

Overall, these applications significantly contribute to a richer, more connected, and enjoyable driving experience. By integrating these diverse services, V2X applications are redefining the concept of travel, transitioning it from a routine task to an enjoyable and productive experience, enhancing both comfort and convenience for all road users.

5. Vehicular communication norms and standards

In this section, we present the norms and standards of communication in V2X [13].

5.1 Dedicated short-range communication

The U.S. Department of Transportation recognizes Dedicated Short Range Communication (DSRC) as a pivotal technology for vehicular communications. DSRC is characterized by its ability to facilitate two-way wireless communications over short to medium ranges, with a strong emphasis on supporting high data transmission rates. These capabilities are especially critical for communication-based security applications within vehicular networks.

Originally developed in the United States under the aegis of the Federal Communications Commission (FCC), DSRC has been specifically tailored to meet the needs of vehicular communications. The system is designed to support a communication range of between 300 and 1000 meters, which is optimal for the types of interactions typical in vehicular environments. In terms of data transfer capabilities,

DSRC boasts data rates ranging from 6 to 27 Mbps, accommodating the high data requirements of modern vehicular applications. Additionally, it is capable of supporting vehicle speeds of up to 190 km/h, ensuring reliable communication even at high speeds.

DSRC operates in the frequency band of 5.850–5.925 GHz, which amounts to a total bandwidth of 75 MHz. This bandwidth is strategically divided into seven channels, each 10 MHz wide, with an additional 5 MHz guard band positioned at the lower end of the spectrum. This allocation and organization of the frequency band are key to DSRC's efficiency, allowing for robust and reliable communication that is essential for the safety and efficiency of vehicular networks. The careful design and standardization of DSRC make it a critical component in the infrastructure of modern intelligent transportation systems.

5.2 IEEE 802.11p

The IEEE 802.11p standard, an initiative by the IEEE working group, was introduced specifically to cater to Wireless Access in Vehicular Environments (WAVE). This standard is a modification of the conventional IEEE 802.11, with enhancements designed to support intelligent transportation systems applications.

One of the key aspects of IEEE 802.11p is its utilization of Enhanced Distributed Channel Access (EDCA), a functionality derived from the IEEE 802.11e standard, which is instrumental in improving the quality of service in vehicular communications. EDCA is particularly effective in prioritizing communication, especially for security-related messages that are deemed high priority. This prioritization is essential in ITS applications where timely delivery of critical information can significantly impact safety and efficiency.

The prioritization mechanism in EDCA works by varying the Arbitration Inter-Frame Spaces (AIFS) and the Contention Windows (CW). By adjusting these parameters, high-priority messages are given precedence over other, less critical messages. This system ensures that important security messages have a higher likelihood of being transmitted promptly, which is crucial for maintaining real-time communication in vehicular networks.

Additionally, IEEE 802.11p structures channel access time by dividing it into repetitive synchronization intervals, each typically lasting 100 milliseconds. This structured approach aids in maintaining regular communication patterns and enhances the overall efficiency and reliability of data transmission in vehicular environments.

Through these adaptations, IEEE 802.11p significantly enhances the capability of vehicular networks to handle the unique demands of ITS applications, ensuring that vehicles can communicate effectively and safely in a variety of scenarios.

5.3 WAVE protocol

The Wireless Access in Vehicular Environment (WAVE) architecture is a comprehensive suite of standards specifically designed to define the communication stack in Vehicle-to-Everything environments. WAVE plays a crucial role in enabling the real-time exchange of traffic messages between Vehicle-to-Vehicle and Vehicle-to-Infrastructure communications. The primary objectives of WAVE in a transportation scenario are to enhance security, improve traffic management, and reduce congestion, thereby facilitating a more efficient and safer driving experience.

In the context of WAVE, Road Side Units within a vehicular network are typically equipped with dual interfaces. One interface is dedicated to the WAVE stack, which is responsible for handling the specific communication requirements of the V2X network. The other interface is designed for external connections, such as Ethernet, which provides access to broader networks like the Internet. This dual-interface approach allows RSUs to effectively manage both internal vehicular communications and external data exchange.

Similarly, On-Board Units in vehicles also feature two types of interfaces under the WAVE architecture. The first is dedicated to the WAVE communication stack, enabling the OBU to participate actively in the V2X communication network. The second interface is used for connecting to various sensors and for facilitating human interaction, such as through user interfaces or display systems within the vehicle. This arrangement in OBUs ensures that they can not only communicate efficiently with other vehicles and infrastructure but also process and interact with data from vehicle sensors and provide information to the driver or passengers.

Overall, the WAVE architecture's design, with its dual-interface approach for both RSUs and OBUs, is integral to the functionality and effectiveness of V2X networks. It ensures that vehicular communication systems are versatile, capable of handling complex interactions within the network while also maintaining connectivity with external systems and user interfaces.

6. Challenges of vehicular networks

The vehicle network faces a large number of challenges in order to provide reliable services [14].

6.1 Data management and storage

The advent of large-scale vehicular networks, potentially comprising millions of vehicles, brings with it the prospect of generating vast quantities of distributed data. This surge in data creation within Vehicle-to-Everything networks presents a unique set of challenges, particularly concerning data management.

One of the primary issues arising from such large-scale networks is the sheer volume of data that needs to be stored and managed efficiently. Vehicular networks, with their multitude of connected vehicles, will produce data at an unprecedented scale. These data can encompass everything from traffic conditions and vehicle performance metrics to environmental data and user preferences. Managing these data effectively is crucial to ensure that it can be used to enhance traffic safety, improve navigation, and facilitate better overall management of transportation systems.

Additionally, the dynamic nature of vehicular networks adds another layer of complexity to data management. The data in V2X networks are not static; they are constantly being updated and changed as vehicles move and interact with their environment. This dynamism requires a flexible and robust data management system capable of handling continuous updates, high-speed data exchanges, and the real-time processing demands of a constantly changing network.

The challenges of managing data in large-scale, dynamic V2X networks thus call for innovative approaches. These might include advanced data processing techniques, cloud computing solutions, edge computing architectures, and sophisticated algorithms for data storage, retrieval, and analysis. The goal is to develop a data management system that is not only scalable to handle the massive influx of data but also agile enough to adapt to the network's dynamic properties. Such a system is key to unlocking the full potential of V2X networks, enabling them to contribute significantly to the evolution of intelligent transportation systems.

6.2 Quality of Service

Quality of Service is fundamentally defined as a collection of requirements that a network must fulfill during the transmission of data packets from their origin to their destination. In the realm of Vehicle-to-Everything communications, ensuring QoS is particularly challenging, despite ongoing efforts to optimize bandwidth and improve message latency.

The challenge of supporting QoS in V2X environments arises from the diverse range of applications these networks support, each with its own unique set of QoS requirements. For instance, real-time applications commonly found within V2X systems are inherently delay-sensitive. These applications, which often include safety-related communication like collision warnings or emergency vehicle notifications, require rapid and reliable data transmission to be effective. Any significant delay in the communication could potentially lead to ineffective warnings or responses, undermining the primary purpose of these applications.

Consequently, ensuring QoS in V2X networks involves not just the optimization of bandwidth to reduce latency but also encompasses a broader set of parameters such as reliability, packet loss, and jitter, which are critical for the smooth functioning of various V2X applications. Different applications may prioritize these parameters differently based on their specific requirements. Therefore, managing QoS in V2X networks requires a nuanced approach that can adapt to the varying needs of different applications while maintaining the overall efficiency and reliability of the network. This adaptability is key to the success of V2X systems in enhancing road safety, traffic efficiency, and driver convenience.

6.3 Routing

Designing dynamic routing protocols for vehicular networks presents a significant challenge, primarily due to the highly mobile and ever-changing nature of these networks. The key objective of these routing protocols is to efficiently disseminate information from one node (vehicle) to another, adapting to the dynamic conditions of the network. However, several critical issues need to be addressed to achieve this effectively [15].

One of the main challenges is minimizing the delays in transmitting information across the network. In the context of V2X, where vehicles are constantly moving and network topologies are frequently changing, ensuring timely delivery of data is crucial, especially for safety-critical applications. Delayed information can be less relevant or even useless, making it imperative to develop routing protocols that can swiftly handle data transfer between nodes.

Another important consideration is the reduction of control overheads. Control overhead refers to the additional network traffic generated by the routing protocol itself (such as route discovery and maintenance messages), which can consume a significant portion of the network's bandwidth. High control overheads can lead to network congestion and reduced efficiency, particularly in dense vehicular

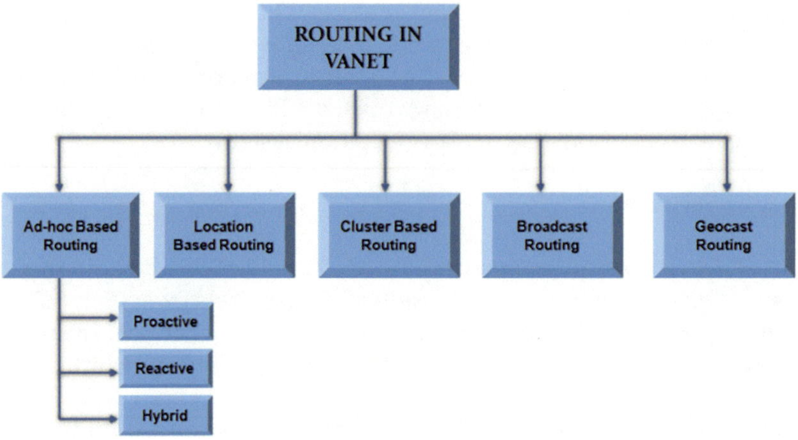

Figure 4.
Routing protocols types in V2X.

environments. Therefore, it is essential to design routing protocols that are not only efficient in data handling but also minimalistic in their own bandwidth usage.

The routing protocol for V2X must be robust enough to cope with the unpredictable and dynamic nature of vehicular network topologies. It should be capable of rapidly adapting to changes such as varying vehicle speeds, different traffic densities, and changing network structures. This requires the protocol to be highly flexible and responsive, with the ability to make real-time adjustments based on current network conditions [16].

Topology-based protocols are classified into three categories: proactive, reactive, and hybrid. For the proactive protocol, the routing table is updated by the nodes by inserting the information of new routes into the network. Hello packets are sent periodically to transfer data to neighboring nodes. This effect creates substantial control overhead and limits the use of available bandwidth. For the reactive approach, updates are sent only when needed, which reduces control overhead substantially. However, this approach still has overhead such as route maintenance. The overhead created in the reactive protocols helps to discover the paths to send the information. Hybrid protocols are considered as a new innovation and discovery made by researchers. This approach focuses on network design architecture more than performance analysis and improvement (**Figure 4**).

In summary, the development of an effective dynamic routing protocol for V2X involves balancing the need for rapid and reliable information dissemination with the constraints of network bandwidth and the dynamic characteristics of V2X. Achieving this balance is key to the successful implementation and performance of vehicular ad-hoc networks.

7. Security in vehicular networks

Security in vehicular networks is crucial because it affects people's lives. In fact, it is essential that transmitted messages cannot be modified or deleted by malicious nodes. Additionally, information about vehicles and their drivers must be secure and protected. Therefore, to ensure the smooth operation of intelligent transportation systems, appropriate security mechanisms must be implemented. In this section, we focus on V2X security in general, presenting the basic aspects and mechanisms of security in V2X [17].

7.1 Characteristics and challenges of V2X security

The characteristics and functionalities of V2X pose certain challenges that may affect the application of security approaches to establish secure communications in V2V and V2I [18].

7.1.1 Mobility

The high mobility of nodes in vehicle networks leads to a highly dynamic network topology, presenting unique challenges in vehicular communication systems. One of the key issues stemming from this dynamic nature is the difficulty nodes face in consistently recognizing their neighbors. This challenge can be exploited by attackers, who may seize the opportunity to create and distribute false topology information within the network.

Attackers, by disseminating inaccurate or misleading information about the network's topology, can manipulate the perceived layout of the vehicular network. This manipulation can lead to the creation of fabricated routes that appear to cross through the attackers' locations. Such deceptive strategies can have severe consequences, as they can be used to orchestrate attacks with the intent of causing traffic disruptions, leading to road congestion or, in more extreme cases, triggering traffic accidents.

These potential attacks highlight the importance of implementing robust security measures in vehicle networks. It is essential to have systems in place that can verify the authenticity of topology information and detect anomalies in data being shared across the network. Mechanisms such as continuous monitoring, data validation, and the use of trusted communication protocols are vital in safeguarding against these types of threats. Additionally, fostering a secure and reliable vehicular communication environment requires collaborative efforts between technology providers, vehicle manufacturers, and regulatory bodies to establish and enforce stringent security standards and practices.

7.1.2 A shared transmission medium

The nature of airborne radio transmission in vehicular networks, where data are transmitted wirelessly, inherently allows for the possibility of passive eavesdropping. An attacker with the right equipment and expertise can exploit this vulnerability by operating in promiscuous mode, a state in which the receiver is able to listen to all data transmitted over the network, not just the data addressed to it. Utilizing sniffing software, the attacker can capture these transmitted packets as they travel through the transmission field.

This capability of passively listening and capturing packets presents a significant security risk. As the process is passive, the attacker does not need to actively interact with or intrude into the network, making the intrusion difficult to detect. The intruder can remain within the transmission field and gather vast amounts of data without the knowledge of the transmitter or other network participants.

Once the data packets are captured, the attacker can attempt to decrypt the transmitted data. Depending on the strength and type of encryption used within the network, this could potentially give the attacker access to sensitive information, such as location data, personal details of the driver or passengers, or other confidential communication.

This scenario underscores the importance of robust security measures in vehicular communication systems. Effective countermeasures include the use of strong encryption algorithms to protect data, implementing secure authentication protocols to verify the identity of network participants, and employing intrusion detection systems to identify and mitigate unauthorized access attempts. Ensuring that these security measures are in place and kept up to date is critical in safeguarding vehicular networks against such passive eavesdropping and data decryption attempts.

7.1.3 Multi-hop communications

The implementation of multi-hop communication protocols in Vehicle-to-Everything networks is crucial for facilitating long-distance wireless communication. In a multi-hop network, data packets are relayed across multiple vehicles (nodes) to reach their intended destination, especially when the source and destination are not within direct communication range. This collaborative nature of multi-hop communication means that the active participation of all vehicles in the network is vital for maintaining its smooth functioning and overall connectivity.

However, the reliance on multiple nodes for data transmission in V2X networks can also introduce security vulnerabilities. Malicious nodes within the network can exploit the multi-hop communication process. As each node potentially acts as a relay for data packets, a malicious node could manipulate or disrupt the data being transmitted. For instance, such a node could alter the content of the packets, delay their transmission, or even create false data, leading to misinformation being spread across the network.

This exploitation can pose serious threats to the security and integrity of the V2X network. It can lead to compromised safety-related information, inaccurate traffic data, or breach of privacy. Therefore, it is imperative to incorporate robust security measures in the design of multi-hop communication protocols for V2X networks. Measures such as strong encryption, authentication of nodes, and implementing intrusion detection systems can help mitigate these risks. Additionally, algorithms designed to identify and isolate malicious nodes can be instrumental in maintaining the integrity and reliability of the network. These security strategies are essential to protect against potential threats and ensure the safe and effective operation of V2X communication networks.

7.1.4 Dissemination of information

In vehicular network protocols, the regular transmission of beacon messages by nodes is a fundamental aspect of maintaining network functionality and ensuring the delivery of various services. These beacon messages typically include information about the node's current position and may contain additional data relevant to specific services being provided within the network. While this practice is essential for the efficient operation of vehicular networks, it also introduces a potential security vulnerability.

Malicious nodes within the network can exploit these regularly transmitted beacon messages for nefarious purposes. By collecting and analyzing these messages, a malicious entity can establish a path profile of specific vehicles. This path profiling involves tracking the movement and behavior of vehicles over time, which can lead to various security and privacy concerns.

For instance, by establishing a path profile, a malicious node can predict the future locations of a vehicle, potentially leading to stalking or targeted attacks. It can also gather sensitive information about a driver's habits, routes, and routines, which could be used for unauthorized surveillance or other invasive activities.

To mitigate these risks, vehicular network protocols need to incorporate robust security measures that protect against such privacy breaches. This might include implementing measures to anonymize beacon messages, limiting the amount of data shared in such messages, or employing advanced encryption techniques to secure the data being transmitted. Additionally, mechanisms to detect and prevent unauthorized data collection and analysis within the network can further safeguard against the exploitation of beacon messages by malicious nodes. Ensuring the privacy and security of users in vehicular networks is critical, and addressing these vulnerabilities is a key aspect of developing resilient and trustworthy vehicular communication systems.

7.2 Entities involved in V2X security

Safety is an issue that must be carefully evaluated and addressed in the design of the vehicular communication system. Several threats potentially exist, including fake messages causing traffic disruptions or even danger, compromising drivers' private information. Issues to be addressed include trust, resilience and effectiveness [19].

7.2.1 The driver

In the V2X safety chain, the driver holds a paramount position due to their crucial role in decision-making and interaction with the system. Despite the advanced technology and automation integrated into V2X applications, the driver remains an indispensable component, particularly in terms of safety.

The driver's omnipresence in the vehicular system is vital for several reasons. Firstly, while V2X technologies provide significant assistance and situational awareness through various applications, it is ultimately the driver who interprets this information and makes real-time decisions based on it. These decisions can range from adjusting speed and changing lanes to more critical responses in emergency situations. The driver's ability to understand and respond to dynamic road conditions, informed by V2X communications, is essential in ensuring overall safety.

Furthermore, the current landscape of V2X applications typically involves the driver as an interactive component within the broader driver assistance systems. These systems are designed to augment, not replace, the driver's capabilities. They provide valuable insights, alerts, and recommendations, which the driver must then evaluate and act upon. For instance, collision avoidance systems may alert the driver of potential hazards, but the driver often has the final say in how to respond to these alerts.

This interactive relationship between the driver and V2X systems underscores the importance of the human element in vehicular safety. It also highlights the need for designing V2X systems that are intuitive and user-friendly, ensuring that the drivers can easily understand and effectively utilize the information and assistance provided to them. As V2X technology continues to evolve, maintaining the driver's central role in the safety chain and ensuring seamless integration between human decision-making and automated systems will be key to maximizing the safety benefits of these advanced vehicular networks.

7.2.2 The vehicle (OBU) and the Road Side Unit

In the context of Vehicle-to-Everything networks, the literature often simplifies the concept of the On-Board Unit by referring to it in terms that encompass both the driver and the vehicle. While this representation does not fully capture the complexity and distinct roles of the driver and the vehicle in the network, it serves as a convenient shorthand for discussing the integrated vehicle system within V2X communications.

Within a V2X network, it is important to recognize two distinct categories of vehicles and Road Side Units:

1. *Normal RSU vehicles and terminals:* these are the standard components of the network and operate as intended. Normal RSU vehicles are equipped with OBUs that facilitate communication with other vehicles and RSUs. The RSUs are strategically placed infrastructure elements that assist in data transmission and network coordination. Both play crucial roles in maintaining the flow of communication and data exchange in the V2X network, contributing to improved traffic efficiency and road safety.

2. *Malicious RSU vehicles and terminals:* contrary to their normal counterparts, malicious RSUs and vehicles represent a security threat within the V2X network. These entities may intentionally disrupt network operations or manipulate data for harmful purposes. Malicious vehicles might spread false information or eavesdrop on network communications, while malicious RSUs could interfere with the data being transmitted between vehicles, leading to misinformation or compromising the network's integrity.

Identifying and mitigating the threats posed by these malicious entities is crucial for maintaining the security and reliability of V2X networks. This involves implementing robust security protocols, continuous monitoring, and advanced authentication mechanisms to ensure that all nodes in the network—both vehicles and RSUs—are functioning correctly and are not compromised. By addressing these security concerns, V2X networks can effectively support their primary goals of enhancing road safety, traffic management, and overall driving experience.

7.2.3 The attacker

In the realm of Vehicle-to-Everything security, the concept of an attacker is broadly defined as one or more entities with the intent to compromise the security and integrity of the network. These attackers employ a variety of techniques to achieve their objectives, which can range from disrupting the normal operation of the network to accessing confidential information or causing physical harm through manipulating vehicular behaviors.

An attacker in a V2X context is not limited to a single entity; it can also be a collective of entities working in concert. For instance, a group of vehicles could collaborate to launch a coordinated attack on the network. This cooperative approach can amplify the impact of the attack, making it more challenging to detect and counteract. Such collaborative attacks can take various forms, including but not limited to:

1. *Spreading false information:* the vehicles may work together to disseminate false or misleading data across the network, potentially leading to hazardous situations on the road.

2. *Jamming or disrupting communications:* by coordinating their actions, these vehicles could more effectively jam or interfere with the communication channels within the V2X network.

3. *Data eavesdropping and privacy breaches:* a group of attackers could collaborate to intercept and decipher data transmissions, leading to unauthorized access to sensitive information.

Addressing the threat posed by such attackers requires a multifaceted approach to V2X security. This includes the deployment of advanced cryptographic techniques, real-time monitoring and intrusion detection systems, and the development of protocols that can quickly isolate and neutralize malicious entities. Additionally, fostering collaboration and information sharing among honest participants in the network can be an effective strategy in identifying and responding to these security threats. Ensuring the security of V2X networks against both individual and collective attackers is critical for maintaining the trust, integrity, and safety of these increasingly essential systems.

7.2.4 Third parties

In the context of intelligent transport systems (ITS), third parties play a significant role as direct participants and facilitators. These third parties typically include entities such as transport regulators, car manufacturers, and traffic police, each of whom plays a distinct role in the overall functioning and management of the ITS.

An important aspect of ensuring secure communication and operations within ITS is the use of cryptographic methods, such as public key pairs. These public keys are essential for establishing secure communication channels, verifying identities, and ensuring the integrity of transmitted data.

1. *Public key pairs:* each of these third parties possesses their own public key pair, which is a part of asymmetric cryptography. The public key is used for encrypting data or verifying a digital signature, while the corresponding private key is kept secure for decrypting data or creating a digital signature.

2. *Integration into OBUs:* these public keys can be integrated into On-Board Units (OBUs) of vehicles. OBUs are envisioned to be tamper-proof devices, which means they are designed to resist unauthorized alterations or extractions of sensitive information, including cryptographic keys. By embedding the public keys of third parties into these devices, secure and authenticated communication can be established between the vehicle and various external entities like regulatory bodies, law enforcement, and manufacturers.

This integration is crucial for several applications within ITS, such as secure Vehicle-to-Infrastructure communication, remote vehicle diagnostics, and enforcement of traffic laws. For instance, a vehicle may need to authenticate a message received from a traffic regulator or send encrypted data to the manufacturer for maintenance purposes.

Overall, the incorporation of public key pairs from third parties into OBUs enhances the security framework of ITS, enabling trusted interactions and protecting against various cyber threats. This is a key component in building a robust and secure intelligent transportation ecosystem.

7.3 Classification of V2X attacks

Like any system and especially regular vehicular networks, V2X is exposed to several types of attacks and threats. In order to protect this network against these attacks, it is necessary to classify them (**Table 1**) [20, 21].

Table 1 presents a detailed overview of various attack types that are critical to the security of Vehicle-to-Everything communication systems. These attacks are grouped into four major categories: attacks on availability, authenticity, confidentiality, and integrity, each targeting a specific aspect of V2X communication security.

Starting with attacks on availability, these are particularly crucial because they target the very essence of V2X systems: ensuring that network services are always operational, and that real-time information is accessible for user safety. In this category, Denial of Service attacks are prominent, aiming to disrupt the availability of network services. Jamming attacks follow, focusing on interfering with communication channels, while Black Hole attacks involve a node that intercepts packets but refuses to participate in routing. Malware attacks, using malicious software, aim to disrupt network operations.

The second category deals with attacks on authenticity, a vital component in ensuring that all network stations correctly authenticate before accessing services. Any breach in this area can lead to severe consequences. This category includes Sybil attacks, where an attacker creates multiple fake identities for a single node; GPS spoofing, which manipulates a vehicle's location data; and Identity Theft attacks, where an attacker impersonates a legitimate vehicle, leading to potential misdirection and misinformation within the network.

Attacks on confidentiality are also covered, highlighting the importance of maintaining the privacy and confidentiality of data within V2X communications. Without

Attack types	Attack description	Attacks and threats
Attacks on availability	Availability is a very important factor for V2X. It ensures that the network and all these services are functional and that useful information is available in real time. This critical security requirement for V2X, whose main purpose is to ensure the lives of users, is an important target for most attackers.	Denial of Service attack Jamming attack Black Hole attack Malware attack
Attacks on authenticity	Authenticity is a major challenge in V2X security. All existing stations on the network must authenticate before accessing available services. Any breach or attack involving the identification or authentication process exposes the entire network to serious consequences.	Sybil attack GPS spoofing Identity Theft attack
Attack on confidentiality	Privacy is an important security requirement for communications in V2X, it ensures that data is only read by authorized entities. In the absence of a mechanism to guarantee the confidentiality of data exchanged between nodes in a vehicular network, the messages exchanged are particularly vulnerable to attacks such as poor collection of clear information.	Eavesdropping attack Traffic analysis attack Man in the Middle attack
Attacks on integrity	The integrity of data exchanged in a system aims to ensure that this data has not been modified in transit. Integrity mechanisms therefore help protect information against modification, deletion, or addition attacks.	Temporal attack Illusion attack

Table 1.
V2X attacks classification.

robust security measures, the exchange of data between nodes is vulnerable to unauthorized access. Eavesdropping attacks involve the interception of data without authorization, Traffic Analysis attacks focus on deducing information from network communication patterns, and Man in the Middle attacks involve an attacker intercepting and potentially altering communications between legitimate parties.

Lastly, **Table 1** discusses attacks on integrity, which are crucial to ensure that data remains unaltered and trustworthy during transmission. This category includes Temporal attacks, where malicious nodes intentionally delay the transmission of messages, rendering them ineffective or even dangerous. Illusion attacks are also featured, where false data is generated through sensors and distributed across the network, leading to potentially misleading information being acted upon by drivers.

In summary, each of these attack types poses a significant risk to the security of V2X networks and, by extension, to the safety and efficiency of the users relying on these systems. **Table 1** serves as a comprehensive guide to understanding the various vulnerabilities and threats that V2X communication faces [22, 23].

7.4 Mechanisms of security in vehicular communications

V2X uses radio transmissions to maintain communication between vehicles, this allows a malicious node to easily inflict itself on the network. In order to deal with these malicious nodes and the attacks previously mentioned, several security mechanisms have been adopted [24, 25].

7.4.1 Tamper-proof device

The tamper-proof device (TPD) is designed to enhance the security of both internal and external communications in vehicles, safeguarding them against unauthorized access and tampering. This device serves a critical function by securely storing sensitive data, such as cryptographic key pairs and digital certificates, and is tasked with the responsibility of digitally signing outgoing messages to ensure their authenticity [26].

To bolster its defense against potential attacks, the TPD is equipped with its own independent power source—a battery that can be recharged using the vehicle's power system. Additionally, it includes a clock that can be securely synchronized at trusted base stations, maintaining accurate timekeeping essential for various security protocols.

Access to the TPD is stringently controlled, restricted solely to authorized personnel. This restricted access is crucial for maintaining the integrity of the device and the data it holds. For instance, the cryptographic keys stored within the TPD can be updated or renewed during the vehicle's periodic technical inspections, ensuring they remain secure and up to date.

A notable feature of the TPD is its array of sensors designed to detect any attempts at physical tampering. In the event that tampering is detected, the device is programmed to erase all stored keys and information. This self-erasure mechanism is a vital security measure, preventing compromised data from falling into the hands of attackers and ensuring the continued security of the vehicle's communication systems [27].

7.4.2 Intrusion detection systems

For effective detection and response to potential attacks on a system, it is essential to have a robust mechanism or software that continuously monitors the data flowing through the system. This mechanism must be adept at identifying and reacting to data that appears suspicious or anomalous. Intrusion detection systems (IDS) are widely regarded as the most effective tools for this purpose [28].

An Intrusion Detection System operates by analyzing network activity, with a focus on identifying patterns or behaviors that deviate from the norm, which could indicate a security breach or attack. Upon detecting such anomalies, the IDS does not just passively report these findings; it actively responds by recommending corrective actions or automatically initiating responses. These responses can range from generating alerts, sending notification emails, to resetting connections to mitigate the threat.

In the context of vehicular networks, IDS plays a crucial role. Here, the IDS is tasked with scrutinizing both incoming and outgoing packets, searching for any indications of malicious activity or signatures. This is particularly important as vehicular networks are highly dynamic and involve a continuous exchange of information, which could be exploited by attackers.

Moreover, IDS solutions are valuable in detecting insider attacks—a type of threat that traditional cryptographic solutions might fail to identify. Insider attacks are perpetrated by entities within the network and often require more sophisticated detection strategies.

Intrusion detection systems can be categorized based on various criteria, such as the detection techniques they employ or their architectural design. The detection techniques might include anomaly-based detection, which looks for deviations from established patterns of normal activity, or signature-based detection, which compares network activities against a database of known attack signatures. Architecturally, IDS can be implemented as network-based systems, monitoring network traffic, or as host-based systems, focusing on activities within individual network hosts [29].

By integrating IDS into vehicular networks, stakeholders can significantly enhance the security and resilience of these systems, safeguarding them against a wide range of cyber threats and ensuring the safety and privacy of their users [30].

8. Conclusion

To conclude this chapter, it is essential to encapsulate the key insights and forward-looking perspectives. This chapter has comprehensively examined the multifaceted aspects of vehicular networking, emphasizing the pivotal role of security in these increasingly connected systems. As the integration of V2X communication in intelligent transportation systems (ITS) becomes more prevalent, the importance of robust security mechanisms to safeguard against a variety of cyber threats becomes paramount.

The exploration of various attack vectors, including threats to availability, authenticity, confidentiality, and integrity, underscores the complexity and evolving nature of security challenges in vehicular networks. These threats not only jeopardize the functionality of the ITS but also pose significant risks to user safety and privacy.

Looking ahead, this chapter suggests that the future of vehicular networking security lies in the development of more advanced and adaptive security protocols, enhanced encryption methods, and innovative intrusion detection systems. The integration of artificial intelligence and machine learning could provide new avenues for predictive security measures, potentially identifying and mitigating threats before they materialize.

DOI: http://dx.doi.org/10.5772/intechopen.114877

Furthermore, as V2X communication becomes more intertwined with other emerging technologies like autonomous vehicles and smart city infrastructure, a holistic approach to security, encompassing all interconnected systems, will be essential. This approach should not only focus on technological solutions but also consider regulatory frameworks, standardization, and collaboration among stakeholders in the automotive, technology, and governmental sectors.

In conclusion, while the advancements in vehicular networking bring forth remarkable opportunities for efficiency, safety, and convenience in transportation, they also introduce complex security challenges. Addressing these challenges effectively will require ongoing research, innovation, and collaboration, ensuring that as vehicles become more connected, they also become more secure, resilient, and trustworthy.

Author details

Mariyam Ouaissa[1*], Mariya Ouaissa[2] and Zakaria Boulouard[3]

1 Laboratory of Information Technologies, Chouaib Doukkali University, El Jadida, Morocco

2 Computer Systems Engineering Laboratory, Cadi Ayyad University, Marrakech, Morocco

3 LIM, Hassan II University, Casablanca, Morocco

*Address all correspondence to: ouaissa.mariyam08@gmail.com

References

[1] Kiela K, Barzdenas V, Jurgo M, Macaitis V, Rafanavicius J, Vasjanov A, et al. Review of V2X–IoT standards and frameworks for ITS applications. Applied Sciences. 2020;**10**(12):4314. DOI: 10.3390/app10124314

[2] Ouaissa M, Houmer M, Ouaissa M. An enhanced authentication protocol based group for vehicular communications over 5G networks. In: Proceedings of the 3rd International Conference on Advanced Communication Technologies and Networking (CommNet); 04-06 September 2020. Marrakech, Morocco: IEEE; 2020. pp. 1-8. DOI: 10.1109/CommNet49926.2020.9199641

[3] Ouaissa M, Ouaissa M, Houmer M, El Hamdani S, Boulouard Z. A secure vehicle to everything (V2X) communication model for intelligent transportation system. In: Computational Intelligence in Recent Communication Networks. Switzerland: Springer International Publishing; 2022. pp. 83-102. DOI: 10.1007/978-3-030-77185-0_6

[4] Soto I, Calderon M, Amador O, Urueña M. A survey on road safety and traffic efficiency vehicular applications based on C-V2X technologies. Vehicular Communications. 2022;**33**:100428. DOI: 10.1016/j.vehcom.2021.100428

[5] Cali U, Kuzlu M, Elma O, Gucluturk OG, Kilic A, Catak FO. Cybersecurity and digital privacy aspects of V2X in the EV charging structure. In: Proceedings of the European Interdisciplinary Cybersecurity Conference (ICC 2023); 14-15 June, 2023. Stavanger, Norway: ACM; 2023. pp. 174-180. DOI: 10.1145/3590777.3591406

[6] Chen S, Hu J, Zhao L, Zhao R, Fang J, Shi Y, et al. The requirements of V2X applications. In: Cellular Vehicle-to-Everything (C-V2X). Singapore: Springer Nature; 2023. pp. 49-79. DOI: 10.1007/978-981-19-5130-5_2

[7] Farag MM, Rakha HA. Development and evaluation of a cellular vehicle-to-everything enabled energy-efficient dynamic routing application. Sensors. 2023;**23**(4):2314. DOI: 10.3390/s23042314

[8] Chen S, Hu J, Zhao L, Zhao R, Fang J, Shi Y, et al. V2X network architecture and standards system. In: Cellular Vehicle-to-Everything (C-V2X). Singapore: Springer Nature; 2023. pp. 81-116. DOI: 10.1007/978-981-19-5130-5_3

[9] Hasan M, Mohan S, Shimizu T, Lu H. Securing vehicle-to-everything (V2X) communication platforms. IEEE Transactions on Intelligent Vehicles. 2020;**5**(4):693-713. DOI: 10.1109/TIV.2020.2987430

[10] Han J, Shi H, Chen L, Li H, Wang X. The car-following model and its applications in the V2X environment: A historical review. Future Internet. 2021;**14**(1):14. DOI: 10.3390/fi14010014

[11] Liu L, Chen C, Pei Q, Maharjan S, Zhang Y. Vehicular edge computing and networking: A survey. Mobile Networks and Applications. 2021;**26**:1145-1168

[12] Shaer I, Haque A, Shami A. Availability-aware multi-component V2X application placement. Vehicular Communications. 2023;**43**:100653. DOI: 10.1016/j.vehcom.2023.100653

[13] Kim S, Shrestha R. V2X current security issues, standards, challenges, use cases, and future trends. In: Automotive Cyber Security. Singapore:

Springer Nature; 2020. pp. 183-212. DOI: 10.1007/978-981-15-8053-6_8

[14] Khezri R, Steen D. Vehicle to everything (V2X)-a survey on standards and operational strategies. In: Proceedings of the 2022 IEEE International Conference on Environment and Electrical Engineering and 2022 IEEE Industrial and Commercial Power Systems Europe (EEEIC/I&CPS Europe); 28 June 2022-01 July 2022. Prague, Czech Republic: IEEE; 2022. pp. 1-6. DOI: 10.1109/EEEIC/ICPSEurope54979.2022.9854561

[15] Chatterjee T, Karmakar R, Kaddoum G, Chattopadhyay S, Chakraborty S. A survey of VANET/V2X routing from the perspective of non-learning-and learning based approaches. IEEE Access. 2022;**10**:23022-23050. DOI: 10.1109/ACCESS.2022.3152767

[16] Tomás VR, García LA, Alonso AL. An agent-based platform to evaluate V2X routing road traffic scenarios. Simulation Modelling Practice and Theory. 2023;**125**:102750. DOI: 10.1016/j.simpat.2023.102750

[17] Yoshizawa T, Singelée D, Muehlberg JT, Delbruel S, Taherkordi A, Hughes D, et al. A survey of security and privacy issues in v2x communication systems. ACM Computing Surveys. 2023;**55**(9):1-36. DOI: 10.1145/3558052

[18] Alnasser A, Sun H, Jiang J. Cyber security challenges and solutions for V2X communications: A survey. Computer Networks. 2019;**151**:52-67. DOI: 10.1016/j.comnet.2018.12.018

[19] Ghosal A, Conti M. Security issues and challenges in V2X: A survey. Computer Networks. 2020;**169**:107093. DOI: 10.1016/j.comnet.2019.107093

[20] Verma A, Saha R, Kumar G, Kim TH. The security perspectives of vehicular networks: A taxonomical analysis of attacks and solutions. Applied Sciences. 2021;**11**(10):4682. DOI: 10.3390/app11104682

[21] Houmer M, Ouaissa M, Ouaissa M. Secure authentication scheme for 5g-based v2x communications. Procedia Computer Science. 2022;**198**:276-281. DOI: 10.1016/j.procs.2021.12.240

[22] Jayatunga E, Nag A, Jurcut AD. Security requirements for vehicle-to-everything (V2X) communications integrated with blockchain. In: Proceedings of 2022 Fourth International Conference on Blockchain Computing and Applications (BCCA); 5-7 September, 2022. San Antonio, TX, USA: IEEE; 2022. pp. 208-213. DOI: 10.1109/BCCA55292.2022.9922372

[23] Ivanov I, Maple C, Watson T, Lee S. Cyber security standards and issues in V2X communications for Internet of Vehicles. In: Living in the Internet of Things: Cybersecurity of the IoT; 28-29 March 2018. Vol. 2018. London, UK: IET; 2018. pp. 1-6. DOI: 10.1049/cp.2018.0046

[24] Yoshizawa T, Preneel B. Survey of security aspect of v2x standards and related issues. In: Proceedings of the IEEE Conference on Standards for Communications and Networking (CSCN); 28-30 October 2019. Granada, Spain: IEEE; 2019. pp. 1-5. DOI: 10.1109/CSCN.2019.8931311

[25] Sedar R, Kalalas C, Vázquez-Gallego F, Alonso L, Alonso-Zarate J. A comprehensive survey of v2x cybersecurity mechanisms and future research paths. IEEE Open Journal of the Communications Society. 2023;**4**:325-391. DOI: 10.1109/OJCOMS.2023.3239115

[26] Rehman MA, Numan M, Tahir H, Rahman U, Khan MW, Iftikhar MZ.

A comprehensive overview of vehicle to everything (V2X) technology for sustainable EV adoption. Journal of Energy Storage. 2023;**74**:109304. DOI: 10.1016/j.est.2023.109304

[27] Tahir MN, Leviäkangas P, Katz M. Connected vehicles: V2V and V2I road weather and traffic communication using cellular technologies. Sensors. 2022;**22**(3):1142. DOI: 10.3390/s22031142

[28] Huang J, Fang D, Qian Y, Hu RQ. Recent advances and challenges in security and privacy for V2X communications. IEEE Open Journal of Vehicular Technology. 2020;**1**:244-266. DOI: 10.1109/OJVT.2020.2999885

[29] Qian Y, Ye F, Chen H-H. Security in V2X communications. In: Security in Wireless Communication Networks. Hoboken NJ: IEEE; 2022. pp. 311-331. DOI: 10.1002/9781119244400.ch15

[30] Lim J, Oh H, Sim K, Kim S, Kim KH. A V2X access authorization mechanism based on decentralized ID (DID) and verifiable credentials (VC). In: 2023 International Conference on Information Networking (ICOIN); 11-14 January 2023. Bangkok, Thailand: IEEE; 2023, 2023. pp. 801-805

Chapter 5

VANET: A Machine Learning Approach

Omkar Pattnaik, Sasmita Pani and Binod Kumar Pattanayak

Abstract

Emerging automotive networks should make the automotive work more secure, greener, and more efficient, and clear the way for autonomous driving to the arrival of the 5G cell. Various new irritants arise due to high sequence factors in vehicle conditions and that is why reconsideration of remote design approaches is also considered. The smart vehicle of the future, which are the backbone of high-performance multi-purpose networks, are gradually being given more advanced sensors and continue to produce information volumes. Machine learning (ML), as an integral part of artificial intelligence (AI), has recently been used in wireless networks to provide continuous information to deal with culturally challenging issues. In this chapter, we examine the recent progress in the use of machine learning in automotive networks and try to pay close attention to this emerging domain. We began to differentiate the obvious characteristics of multi-functional vehicle networks and promote the use of a learning machine to deal with subsequent difficulties.

Keywords: VANET, machine learning (ML), artificial intelligence (AI), autonomous driving, multi functional

1. Introduction

Despite the fact that VANET is derived from the concept of MANET, there are some subtle contrasts between them. A portable remote device is used to communicate between nodes in a MANET, but in the case of a VANET, running vehicles talk to other vehicles. VANETs are made of vehicles and highway guides. It is a critical method of a competent moving framework. The development of nodes in VANET is traditional as vehicles move along established roads with high momentum limited to the bottom of situations such as small or curved street, congestion/congestion and so on.

In designing this environment, we assume that each vehicle must be associated with an on-board unit (OBU) and a road-side unit (RSU) installed along the roads. RSUs and OBUs are required to communicate and exchange information with each other through a protocol called Dedicated Short-Range Communications (DSRC) [1]. Three types of communication are possible in VANET such as Vehicle to Vehicle (V2V), Vehicle to Infrastructure (V2I) and Infrastructure to Infrastructure (I2I). But V2V method uses DSRC protocol. It is maintained purely as an ad hoc network. The main goal of DSRC is to ensure low communication delay and high data transmission in the network [2]. The Federal Communications Commission (FCC) subsequently updated the DSRC protocol to WAVE (Wireless Access for Vehicular Environment).

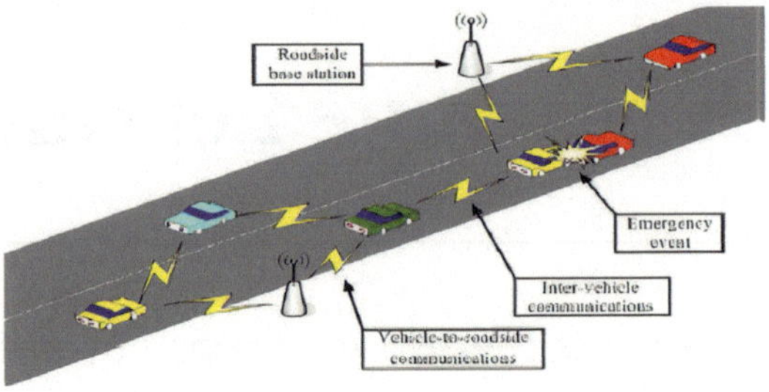

Figure 1.
Structure of VANET.

The pattern of VANET Structure is given in **Figure 1**.

Remote network which can support large portability broadband get to have gotten increasingly more consideration from both industry and the scholarly community as of late [3, 4]. Specifically, the idea of associated vehicles or vehicular systems, as appeared in **Figure 2**, has increased considerable energy to bring another level of availability to vehicles and, long side novel onboard figuring and detecting advancements, fill in as a key empowering influence of Intelligent Transportation System (ITS) and developed urban areas [5]. This new age of networks will at last have a significant effect on the society, making ordinary traveling more secure, greener, and progressively productive and comfortable. Alongside recent advances in a wide scope of artificial intelligence (AI) advances, it is helping clear the way to autonomous driving in the appearance of the fifth generation cell frameworks (5G).

Meanwhile, scheduled inventive vehicles are progressively furnished with a broad assortment of sensors, for example, "engine control units, radar, light detection and ranging (LIDAR), and cameras" to enable the vehicle to see the general climate just as observing its individual activity quality in current world. Along with superior concluding and capacity gadgets onboard, these sensible innovations are switching vehicles from a basic remodel efficiency to a capable figuring and connections centre with rational handling potentiality. They continue gathering, producing, putting away, and conveying large volumes of information, subject to additionally processing

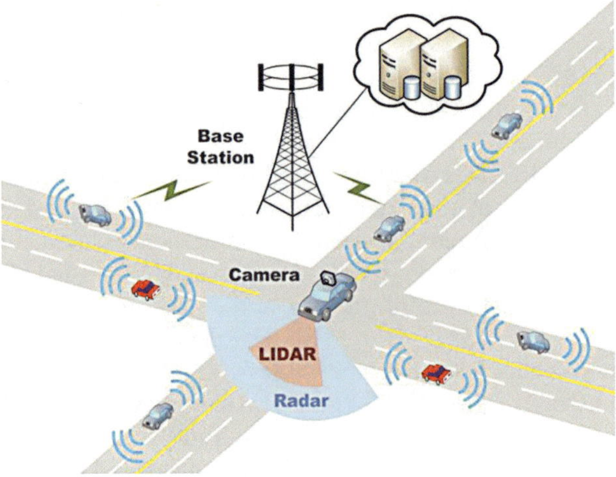

Figure 2.
Structure of VANET with LIDAR.

and usually called as mobile big data [6]. Such information give rich context data in regard to the vehicle energy, (for example, speed, acceleration and direction), street conditions, traffic flow, wireless environments, and so on that can be motivated to improve network performance through versatile data driven dynamic. Generally, traditional interchanges techniques are not intended to deal with and exploit such data.

The applications produced for VANETs can be grouped into 3 principle classes: "(1) Safety applications (e.g., road risk control warning and emergency electronic break light), (2) comfort applications (e.g., parking accessibility notice and blocked road notice), and (3) commercial applications (e.g., service announcements and content map database download) [7]." These programs produce two types of messages for VANET exchanges, which include secure information and non-secure information. Safety information, including reference points and urgency information, travel on the control path. Non-secure information, including information generated by convenience or business applications, are routed to the favorable path [8, 9].

In city conditions, the junctions are imperative locations. So these are the often probably positions for traffic accident. In "Canada, nearly 800 avenue customers killed and 7250 critically injured at intersection clashes in 2005 [10]." The latest statistical information of avenue clashes mentioned via "Canada Road Safety Vision 2010 indicates that approximately 47% of all human beings killed and 57% of all injured [3]." Moreover, primarily located on the collective facts of avenue clashes in "Brampton roadways, between 2003 and 2007, 71% of pedestrian collisions came about at intersections [3]." To furnish safety and greater reliable surroundings about road customers on the junction of city roads, the use of VANETs appear to be crucial.

An eminent level of "Quality of Service (QoS)" is essential at junctions to prevent any verbal exchange conflicts that might also appear due to fleshy conversation bundle. Development "of QoS in VANETs" is a demanding assignment due to may specific VANET features such as topology settings and common damaged paths [11].

2. Related work

Machine learning methods can gain from a dataset via naturally breaking down the dataset to recognize rules. We can utilize these principles to anticipate results from new information. When all is said in done, machine learning frameworks need a pretraining procedure with the end goal that they can generalize input information. As indicated by various training information, machine learning procedures can be arranged as either unsupervised clustering or supervised classification. Unsupervised machine learning frameworks analyze the similarity between information to choose whether such information can be gathered. Supervised machine learning frameworks map input information to the ideal outputs, characterizing input information as per their qualities.

Currently machine learning frameworks have been utilized to improve the presentation of wireless networks. When packets are lost in the traditional TCP-accommodating rate control protocol, it is viewed as network congestion. The protocol will decrease the bundle transmission rate and lower channel usage. In Ref. [12], the authors use machine figuring out how to decide the reason for packets misfortunes and take various measures to improve the system execution. The reason can be congestion, a path change, or connection errors. The cluster-based routing rules can assemble nodes into clusters and recognize cluster heads and general hubs through machine learning procedures [13]. In Ref. [14] communicate routing utilizes machine learning to foresee whether a packet should be rebroadcast. In Ref. [15], the authors powerfully alter the reference point interval through machine learning, which diminishes the control overhead and keeps up the unwavering quality of transmissions.

With regard to vehicular correspondences, the authors explain information communicated over the system originates from various sensors and On-Board Units (OBUs). Mischief recognition, explicitly information driven trouble making identification, is a significant piece of the security framework. It centres on choosing the accuracy of information communicated and contrasting it and past data received. The greater part of the work directed in information driven either requires agreement [16] or the presence of a central authority [17]. Introduced an adaptable central misbehavior assessment framework for OBUs and roadside infrastructure units that identifies and dispenses with aggressors from the system dependent on information believability. The methodology included each node, recognizing an incident, to advance report to the central power to take a ultimate choice.

K-Nearest Neighbor (K-NN): K-NN evaluation is a common anomaly detection/ classification technique and broadly used due to its easy implementation and capability to naturally cope with multiclass cases [18].

In this case, authors have mentioned regular K-NN which analyses the conduct of statistics factors in accordance to its previously broadcasted role primarily based on "Euclidean distance." "A data point is labelled as an outlier if it is situated farther away from its earlier broadcasted points. In Third International Knowledge Discovery and Data Mining Tools Competition, KDD cup 1999 dataset was once used to build a IDS the use of classification algorithms. The 1-nearest neighbour classifier algorithm, 1-NN, performed similar to the winning entry and proven magnificent results in terms of detection" [19].

"Li et al. [20] investigated the effect of the Flooding Attack in wi-fi sensor networks and introduced an intrusion detection device based on K-NN algorithm. The K-NN algorithm" was correctly in a position to observe the ordinary nodes using the frequency of sent messages as a characteristic vector.

Support Vector Machine (SVM): In Ref. [6] the authors trying to focus on SVM is a discriminative classifier, characterized by an isolating hyper plane, that maps the information focuses on the plane. This direct machine has a place with the class of Kernel techniques with issue explicit portion work. The job of Kernel work in "SVMs is to prompt an element space by mapping the training information on a high dimensional space where it is straight distinct. SVM have shown achievement in the field of text mining, pattern recognition, object classification, and anomaly detection. Ref. [10] displayed classification" of various kinds of interruptions utilizing supervised learning procedures on the ISC2012 dataset, a well-known benchmark dataset for interruption location [2]. The work exhibited that "SVM, with a discovery precision of 99.1%, could be utilized effectively as an classification module of a network anomaly IDS".

"In Ref. [21], the author showed a cloud-based mobile distributed intrusion detection system for VANET that detects Denial-of-Service attack using One Class SVM (OCSVM) classifier with RBF kernel and score the peculiarities utilizing recursive K-implies clustering. The framework exactness was compared on the quantity of transmitted reference point by vehicles, the speed of intruder vehicle, and the separation between Road Side Units".

3. Vehicular networks facing difficulties with high mobility factor

High portability vehicular systems show unmistakable qualities, which have presented critical "provocation to wireless network" structure. In current segment, we recognize alike difficulties and afterward examine the capability of utilizing "machine learning" into the address domain.

3.1 Stable mobility

"Higher flexibility of vehicles leads to stable dynamics and affects machine design in multiple aspects of the connections network. One of the most important ways that high mobility sets itself apart is through its unique channel propagation component networks compared as low mobility counterparts. For example, vehicular channels exhibit rapid sensual variation and also suffer from inherent non-stationary of carrier statistics as on their different natural environment effective" [7, 22]. "This one is made worse by the fact that carrier statistics, which are typically used to increase evaluation accuracy, are non-stationary" [8].

"Meanwhile, due to then high Doppler spread caused by vehicle mobility, then multicarrier modulation scheme is maximum susceptible into inter carrier interference (ICI) in vehicular networks" [10, 11] and hence brings difficulty to signal identification. Constant movement of vehicles also causes usual modifies of then communications network arrangement, affecting channel allocation and routing protocol designs. For instance, in cluster-based vehicular networks [12], "moving vehicles may combine and leave the cluster repeatedly, making it hard to maintain long-lasting communications within then formed cluster and so warranting further study on cluster stability. Another sources of dynamics in higher flexibility networks comes from then unstable vehicle density, which modify dramatically reliant on then locations" and moment "(peak or off hours of the day)". "Flexible and robust resources management plan that makes an efficient use to available resources while adapting to then vehicle density changing are thus required".

3.2 Independent and demanding QoS provisions

In large portability vehicular systems, there may be chances of various sorts of links, that we comprehensively order "into V2I and V2V connections. The V2I" joins empower vehicles to convey "with the base station to help different traffic efficiency and data (infotainment) services". "They generally require visit access to the Web or remote servers for media streaming, map downloading, and social" communication, that include impressive measure of information convection and in this way are maximum transfer speed concentrated [5]. Then again, the V2V joins are for the most part considered for sharing safety basic data, as basic safety messages (BSM) in DSRC [23], among vehicles in close proximity in either an intermittent or occasion activated way. Such safety related messages are carefully defer touchy and require high unwavering quality.

Specifically, "reinforcement learning [24], one of the machine learning devices, can collaborate with the dynamic" condition and create palatable customs to satisfies differing QoS prerequisites about vehicular systems during adjusting to shifting mobile condition. For instance, "in resource allocation problems, the optimal strategies are first learned and afterward the vehicle agents" in like manner catch activities to modify controls and appropriate channels versatile to the changing situations portrayed by, e.g., link conditions, privately saw obstruction, and vehicle energy while conventional static numerical models are not so good at catching and tracking such powerful changes [25, 26].

4. Introduction to machine learning

Machine learning (ML) permits computers to discover concealed bits of knowledge through iteratively gaining from information, without being expressly customized. It has changed the universe of computer science by permitting learning with enormous datasets, which empowers machines to change, re-structure and advance algorithms without anyone else. Existing machine learning strategies can be separated into 3 classes, to be specific, supervised learning, unsupervised learning, and reinforcement learning. Other learning plans, for example, semi-administered learning, web-based learning, and transfer learning, can be seen as variations of these three essential sorts. When all is said in done, ML includes 2 phases, i.e., training and evaluating. "In the training stage, a model is found out dependent on the training" information "while in the testing stage, the trained model is applied" to create the expectation. In current segment, we quickly present the idea of" ML with the expectation which then researchers can value may be in taking care of generally testing issues (**Figure 3**).

4.1 Supervised learning technique

The popularity of practical ML algorithms utilize supervised learning with a labeled dataset, where each preparation test accompanies a label. A definitive objective "of supervised learning is to discover the mapping from the input data space to the label" with the goal that dependable expectation could be done when fresh knowledge is provided. Issues with supervised learning can also be divided into "classification and regression," in which the contrast the difference between the two projects is that the labels for classification are absolute while those for regression are numerical. Allocation procedures figure out how to anticipate a classification output for every approaching sample dependent on the training information. Some classic procedures in this category incorporate Bayesian classifiers [15], k-nearest neighbors (KNN) [27], decision trees [14], support vector machine (SVM) [16], and neural networks [28]. Rather than discrete outputs, regression procedures predict a consistent value relating to each sample, for example, estimate element inputs. Classic regression procedures incorporate logistic regression [24], support vector regression (SVR) [17], and Gaussian process for regression [29].

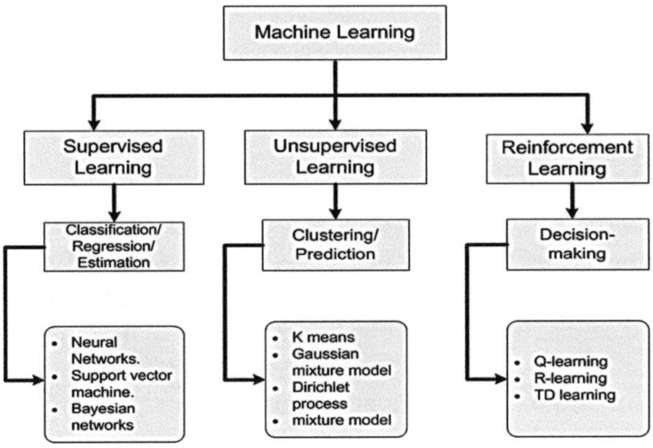

Figure 3.
Classification of machine learning structure.

"If the training data only includes discrete values then it is a classification problem and the output of the trained model is a classification which is also discrete". "On the other hand if the training data containing continuous values, then it is the regression problem and the output of the trained model will be a prediction. Two widely used examples of supervised ML are Decision Trees and Random Forest".

The result of reverting procedures is a constant worth that may represent to forecast of the house value, stock trade, banking client exchanges, electric vehicle battery, level of traffic blockage at different crossing points, and jamming expectation (**Figure 4**).

In vehicular social network, regression can be utilized to foresee parameters, for example, network throughput. Two classic regression procedures are:1) Gaussian Process for Regression (GPR) and 2) Support Vector Regression (SVR).

In vehicular systems the classification procedures can be utilized for interruption or malfunction detection. Besides, classification procedures are additionally beneficial in the traffic secure applications, for example, Augmented Reality Head Up Display (AR-HUD), dynamic driver data frameworks, obstacle recognition, and anticipating complex traffic types.

4.2 Unsupervised learning technique

In the unsupervised ML, training depends on unlabelled information. "This plan trying to find an efficient portrayal of the unlabelled information. For instance, the highlights of data can be caught by various hidden factors that can be spoken to by the Bayesian learning methods". Clustering is a kind of unsupervised "discovering that gatherings tests with comparable highlights". The input information of each data point can be its supreme portrayal or a relatives similarity level into other information points (**Figure 5**).

In the wireless methods universally, the cluster organization for the hierarchical protocols is important with respect to energy management, where every node simply needs to speak with the cluster head before speaking with the individuals from different clusters. Some traditional clustering methods are k-means, spectrum clustering, and hierarchical clustering. Measurement reduction is another subclass of unsupervised ML conspire. The principle thought behind measurement reduction is to down-sample the information from a higher measurement to a lower measurement with no significant information loss [19]. Implementing machine learning for most applications require measurement reduction because of various reasons.

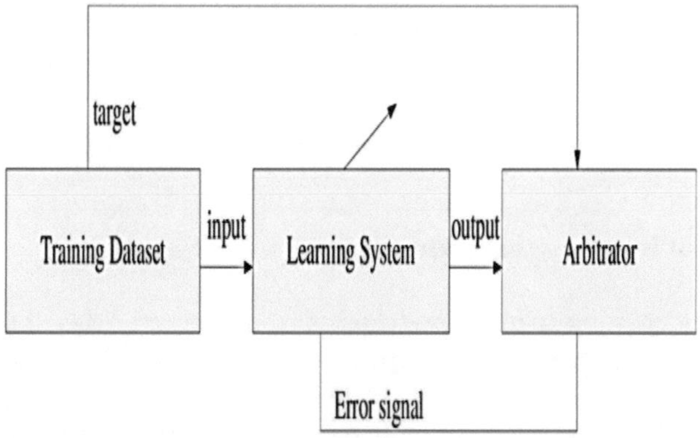

Figure 4.
Supervised learning structure.

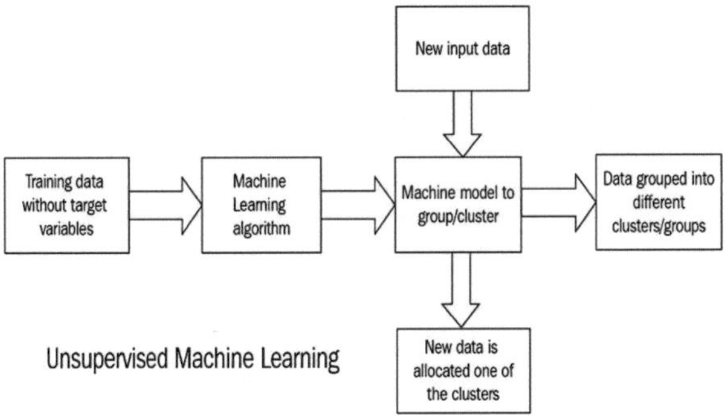

Figure 5.
Unsupervised learning structure.

A delegate instance of unsupervised learning is clustering, to be specific, to assemble tests such that samples in a similar cluster have a larger number of similarities than the samples in various clusters. The highlights utilized for clustering could be either the outright portrayal of each sample or the relative similarities between samples [20].

Other than clustering, measurement reduction is another significant instance of unsupervised learning, where tests from a high dimensional space are anticipated into a lower one without losing an excess of data. In numerous situations, the raw information accompany high measurement, which is not possible due to a few reasons. One explanation is the purported revile of dimensionality [18], which depicts the hazardous marvel experienced when the measurement gets larger. For example, in enhancement, clustering, and classification, the model unpredictability and the quantity of required training samples develop significantly with the element measurement.

Reapproach information dimensionality is the first reason. In clustering, classification and advancement, the ordinary model unpredictability increasing significantly within the expansion of highlight instruments. The second reason is the obstacle in the learning procedure. In the greater part of the cases the highlights of the information samples are correlated in certain angles, yet on that the component value is changed by noise or impedance, at thats position the separate outcome of the "correlation will be depraved and the learning procedure will be influenced. Such type of measurement decreasing in the vehicular informal network is the development which prompts a vehicular cluster". Cluster head gathers and transform the data to the node B to diminish the correspondence costs. The curse of dimensionality can be decreased by the measurement decrement techniques. Measurement reduction strategies are gathered in two classes: "(1) linear projection methods such as: Principal Component analysis (PCA) and Singular Value Decomposition (SVD), and (2) nonlinear projection methods such as: manifold learning, local linear embedding (LLE) and is metering mapping" [21].

4.3 Reinforcement learning technique

Reinforcement learning in effectively gains from the activities of the taking in operator from the relating reward. It implies so as to expand the reward, inexplicit mapping the circumstances as per the activities by experimentation. The Markov Decision Process (MDP) is a case of reinforcement learning. Q function model-free learning process is an exemplary guide to take care of MDP advancement problem that does not require data about the learning condition.

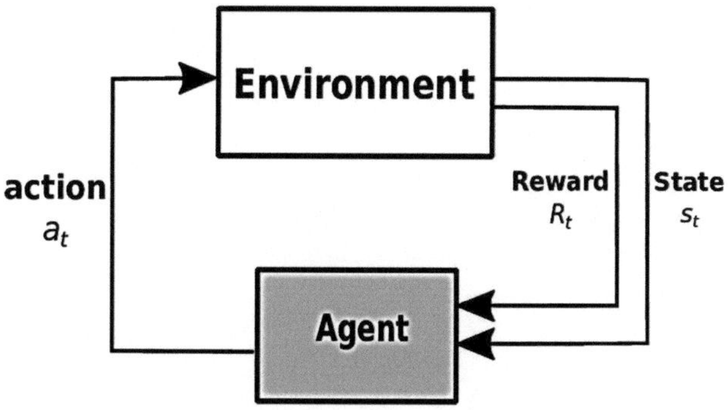

Figure 6.
Reinforcement learning structure.

In reinforcement learning issues, an operator learns the ideal practices through interfacing with nature in an experimentation way intending to augment compensations from nature. The environment is displayed as a Markov decision process (MDP), which acquaints activities and rewards with a Markov procedure. Both the state progress probability (**Figure 6**) [30].

Activities and their prizes produce strategies of the decision of an appropriate activity. In a given express Q work assesses the mean of the aggregate reward. "The perfect Q work is the greatest expected total reward that can be accomplished by following any of the arrangements". Reinforcement leaning is an ideal possibility for tending to different research difficulties in vehicular systems. For instance, helpful advancement of fuel utilization for a "given geographical region, taking care of the spatial and temporal varieties" of the V2V correspondences, ideal path expectation of electric vehicles, and decrease in traffic blockages.

4.4 Deep learning technique

"Deep learning is firmly identified with the above 3 classifications of ML". It is a more deeper system of neurons in different phases. It intends to extract information from the information expressions, that can be produced from the recently examined 3 classes of ML. The system comprises of input layer, hidden layers and an output layer. Every neuron has a non linear transformation function, such as ReLU, sigmoid, and flawed ReLU. The scaling of input information is critical as this can seriously influence the expectation or classification of a system. As the quantity of hidden layers builds, the capacity of the system to adapt additionally increments. However, after a specific point, any expansion in the hidden layers gives no improvement in the exhibition.

The training of a more deep system is additionally challenging since it requires broad computational resources, and the slopes of the systems may detonate or evaporate. The arrangement of these resources hungry more deeper systems has raised the significance of edge computing innovation. Vehicles progressing can get benefit from versatile edge computing servers. **Figure 7** illustrates the idea of neurons, input layer, hidden layers and output layer in a Deep neural network. It is also known as Multi-Layered Perception (MLP).

Deep learning intends to learn information portrayals, which can be worked in supervised, unsupervised, and reinforcement learning and has made huge advances in different machine learning assignments. As a deeper variant of neural systems,

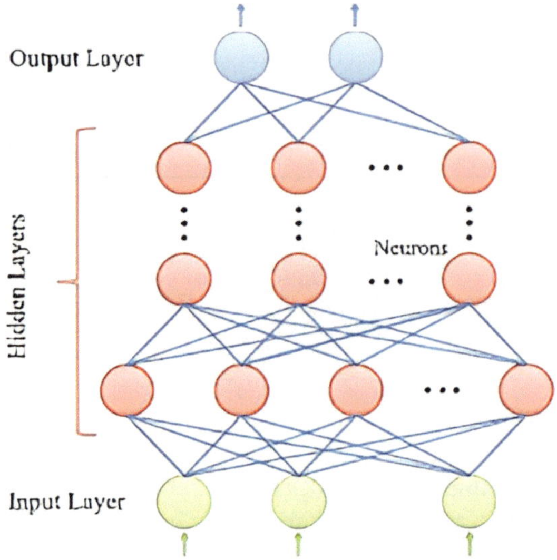

Figure 7.
Deep learning structure with multiple layers.

which comprise various layers of neurons, the structure of deep learning is appeared in **Figure 7**. The input layer is at the left, where every hub in the figure speaks to a component of the information, while the output layer is at the right, relating to the outputs. The layers in the centre are called hidden layers [31].

4.5 Unsupervised learning application in VANET

Machine learning (ML) procedures are strong methods for classifying and clustering the enormous datasets because of their particular capacities, for example, short processing time, supporting the large amount of information, automatically detecting pattern in data, anticipating future data utilizing the uncovered patterns, and intending to gather more information [32]. In this survey, the ML algorithms are utilized for clustering the huge informational set and multi-dimensional highlights space of VANETs. Generally, the machine learning algorithms can be divided into supervised (classification) and unsupervised (clustering) learning algorithms categories [33]. In this survey we have focused on clustering technique in VANET which is the part of unsupervised learning method.

The supervised algorithms that are utilized for labeled information need to utilize a training informational set for looking at the highlights. However, the unsupervised algorithms that are utilized for unlabelled information do not have to utilize the training informational set [34]. "In this survey, the unsupervised machine learning algorithms are utilized for information clustering because of assorted variety of messages over V2V and V2I correspondences and unlabelled information in the VANETs".

"In current survey, then K-implies procedure is utilized as clustering then information in VANETs. Then K-implies is one of the most well-known unsupervised learning procedures broadly utilized as the multi-dimensional highlights clustering. The K-implies is an adaptable, fast and basic learning procedure. Additionally, this procedure is effective as preparing long information. In the specified, the K-implies research procedure utilized in the paper is depicted in major particular".

K-implies procedure was at first presented "by James Macqueen in 1967" [27], yet its smooth one of the better mainstream "unsupervised" clustering procedures because of its effortlessness, experimental achievement, proficiency, and simplicity of usage. K-implies clusters a set of information into k number of clusters dependent on their highlights. For every data, K-implies computes the Euclidean distance to all cluster centroids (cluster centers), and afterward chooses the small distance. The information has a place with a cluster that the distance to the centroid of that cluster is the less. At that point, the new centroid is determined for each cluster dependent on the average coordinate of all members from each cluster. At last, all the information are clustered asked on the new centroid. K-implies repeat this procedure until the individuals from each cluster do not move to different clusters any longer [35, 36].

Essentially, K-implies comprises of 3 fundamental advances: "(1) selecting existing centroids as k clusters; (2) computing squared Euclidean distance of every data to the centroids; and (3) computing then new centroids cluster to discover nearest centroids. Stages 2 or 3 ought to be rehashed unroll the cluster individuals no more change [27, 29]."

The underlying centroids as K-means can be picked by Forgy and Arbitrary techniques [37]. In any case, there is no assurance that, utilizing these strategies, K-implies combines [37, 38]. In this way, the researchers utilize different techniques for deciding the underlying centroids. "In this survey, the underlying centroids for k clusters are thought to be the main k messages got by RSUs."

"K-implies has 3 inputs sources including highlights, number of clusters and number of iterations Clustering" procedures characterize a set of articles dependent on recognized highlights, along these lines, significantly affect execution of "the clustering procedures. In K-implies, the highlights must be changed to the grater qualities. Ordinarily, there is no proficient system as deciding the looks". In fact, the highlights ought to be resolved explicitly as every issue dependent on the information about the area of issues [27]. In the survey, the highlights of K-implies are characterized dependent on the highlights of "information including the message size, validity of messages, span between vehicles and RSUs, type of information and direction of information sender".

"The quantity of clusters are the alternative input as K-implies clustering procedure. The best number of cluster as every issue may be characterized by executing the clustering procedure as various quantities of clusters [27]. In current survey, the quantity of clusters (k) for K-implies is captured by leading a set of fundamental simulations".

The clustering procedure is exited when a predefined merging is produced. That explains there are not any modifies in then clusters' individuals. However, if the merging is not received in the considerable duration, the clustering procedure must be exit after a predefined number of loops. Theoretically, K-implies does not fastly assemble notable for the large information sets [27]. "In current survey, the number of loops is pretended to be 100".

The multifaceted nature of K-implies relies upon then quantity of information in every datum set, the quantity of highlights, number of groups and cycles. In this way, legitimate beginning conditions can bring about a superior bunching [32].

5. Outcome formulation in VANET with learning-based methods

The rich wellsprings of information produced and put away in vehicular systems inspire an information driven methodology for decision making that is versatile to arrange elements and strong to different disabilities. Machine learning speaks to a successful apparatus to fill such needs with demonstrated great execution in a wide assortment of utilizations.

5.1 Network congestion management

Information traffic congestion is a significant issue in vehicular systems, particularly when the system conditions are profoundly dense, e.g., occupied crossing points and swarmed urban situations. In such cases, more number vehicles are competing for the accessible correspondence channels at the same time and thus cause serious information impacts with expanded packet misfortune and delay. To ensure a dependable and convenient conveyance of different deferral sensitive safety basic messages, for example, BSMs, the vehicular systems need to have carefully structured procedure rate-based, power based, carrier-sense multiple access/collision avoidance based, prioritizing and scheduling-based, and hybrid strategies [39], which alter correspondences parameters, for example, transmission power, transmission rates, and dispute window sizes, and so forth, to meet the congestion control purposes (**Figure 8**).

Unique in relation to the traditional methodologies, a viable machine learning based on information traffic jam regulation procedure using k-implies clustering has been created in Ref. [29] for traffic jam inclined convergences. The expected methodology depends on nearby "road side units (RSUs)" introduced at every convergence for traffic jam discovery, data "purification, and traffic jam regulation to arrange a subjective traffic jam authority to every vehicles that are going concluded or close at the" crossing point. Afterwards discovery of traffic jam, all RSU gathers all information moved among vehicles in their inclusion, expels its repetition, misuses k-implies procedures into cluster then informations as indicated by its highlights, for example, content, gravity and groups, lastly changes interchanges parameters for each cluster.

5.2 Security in network

As clever vehicles become increasingly associated and carry tremendous advantages to the general public, the improved availability can make vehicles progressively vulnerable against digital physical attacks. Subsequently, security of data partaking in vehicles are significant afterward any broken sensor estimations may source of hazards and wounds. In this chapter "intrusion detection system" has suggested to vehicular systems dependent on profound "neural networks", point

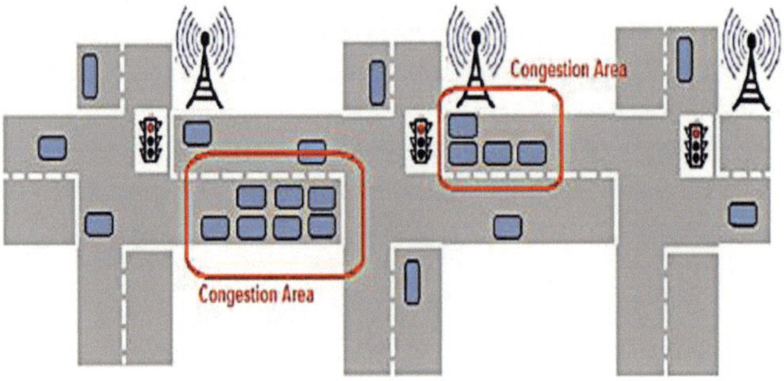

Figure 8.
Intersection of congestion area and RSUs.

the unsupervised rooted profound conviction systems are utilized into introduce the confines as pre preparing phase. At that point, "the profound neural networks are prepared by large-spatial packet information to make sense of the hidden measurable properties of typical and disfigure packets" and concentrate the relating highlights. Moreover, LSTM is utilized in to recognize attacks on associated vehicles. The LSTM based locator can perceive the incorporated abnormalities with high accuracy by figuring out how to anticipate the following word beginning from every vehicle.

5.3 Location forecasting based set up and routing

We are appeared in Segment IV that machine learning may be utilized to become familiar with the elements in higher versatility vehicular systems, along with vehicle direction forecast. In reality, the anticipated elements can be additionally utilized close to systems contract structures for framework execution enhancement. For analysis, the "Hidden Markov model (HMM)" has been enforced in to anticipate vehicles' forthcoming areas dependent on previous portability "follows and development designs in a mixed VANET along with V2I and V2V" joins. In light of the anticipated vehicle directions, a successful routing plan has been proposed to proficiently choose hand-off hubs for message sending and empower consistent handoff somewhere in the range of V2V and V2I correspondences.

"A variable-request Markov model has been received to separate vehicular versatility designs from genuine follow information in a urban vehicular system condition, which is utilized to anticipate the potential directions of moving vehicles and create productive expectation based delicate steering conventions." "In Ref. [28], a recursive least squares procedures has been utilized for enormous scope channel forecast dependent on the location data of vehicles and encourage the improvement of a novel planning procedure for helpful information dispersal in VANETs."

6. Conclusion and future scope

In this chapter, we have explored the chance of applying machine learning to address issues in high versatility vehicular systems. Solid elements displayed by such kinds of systems and the requesting QoS necessities challenge the cutting-edge interchanges innovations. Machine learning is accepted to be a promising answer for this test because of its striking execution in different computer-based intelligence related dimensions. We have analyzed the fundamentals of machine learning and afterward gave a few instances of utilizing such tools to get familiar with the elements and perform intelligent and dynamic in vehicular systems. We have additionally featured some territories that require more consideration.

This article gives a review of applying machine learning to address difficulties in developing vehicular systems. We have focused the basics of machine learning, including significant classes and agent methods. In this work examples of some models which are applying in machine learning in vehicular systems to encourage information driven dynamic, and discussed intelligent wireless resource management.

We have discussed different artificial intelligence procedures; all advancement issues face vulnerability in the intensions of the encompassing members. Various parts of artificial intelligence can help each other to draw out an ideal arrangement that would not cause or produce issues in the area they are not planned.

The machine learning procedures as a rule require higher computational assets. These assets may not require being inside the vehicle.

To for all intents and purposes to implement the review system in real vehicular systems, a RSU should be set at every crossing point. Likewise, since congestion control is a continuous procedure, RSUs may should be outfitted with Graphic Processing Units (GPUs) for rapidly executing machine learning procedures. We have to remember that, the machine learning procedures direct an enormous number of procedures and activities that takes a great deal of time.

Author details

Omkar Pattnaik[1*], Sasmita Pani[1*] and Binod Kumar Pattanayak[2*]

1 Department of Computer Science and Engineering, Government College of Engineering, Keonjhar, Odisha, India

2 Department of Computer Science and Engineering, Institute of Technical Education and Research, Siksha 'O' Anusandhan Deemed to be University, Bhubaneswar, Odisha, India

*Address all correspondence to: omkar29in@gmail.com; susmitap20@gmail.com and binodpattanayak@soa.ac.in

References

[1] Pattnaik O, Pattanayak BK. Security in vehicular ad hoc network based on intrusion detection system. American Journal of Applied Sciences. 2014;**11**(2):337-346

[2] Pattnaik O, Pattanayak BK. Reliable and faster data communication in MANET and VANET based on localization and clustering techniques. International Journal of Engineering & Technology. 2018;7:66-72

[3] Liang L, Peng H, Li GY, Shen X. Vehicular communications: A physical layer perspective. IEEE Transactions on Vehicular Technology. 2017;**66**(12):10647-10659

[4] Peng H, Liang L, Shen X, Li GY. Vehicular communications: A network layer perspective. IEEE Transactions on Vehicular Technology. 2017. pp. 1-32 Available from: https://arxiv.org/abs/1707.09972

[5] Cheng X, Yang L, Shen X. D2D for intelligent transportation systems : A feasibility study. IEEE Transactions on Intelligent Transportation Systems. 2015;**16**(4):1784-1793

[6] Cheng X, Fang L, Hong X, Yang L. Exploiting mobile big data: Sources, features, and applications. IEEE Network. 2017;**31**(1):72-79

[7] Jabbarpour Sattari M, Noor RM, Ghahremani S. Dynamic congestion control algorithm for vehicular ad hoc networks. International Journal of Software Engineering and Its Applications. 2013;7(3):95-108

[8] Liu Y, Dion F, Biswas S. Dedicated short-range wireless communications for intelligent transportation system applications: State of the art. Transportation Research Record. 2005;**1910**:29-37

[9] Qian Y, Moayeri N. Design of secure and application oriented VANETs. In: Proceedings of IEEE VTC 2008 Spring. 2008. pp. 2794-2799

[10] Pedestrian Safety Plan. 2010. Available from: https://www.brampton.ca/EN/residents/Roads/pedestrian-driver-safety/Documents/Pedestrian%20Safety%20Plan%20-%20Final.pdf

[11] Taherkhani N, Pierre S. Improving dynamic and distributed congestion control in vehicular ad hoc networks. Ad Hoc Networks. 2015;**33**:112-125

[12] Grover J et al. Machine learning approach for multiple misbehavior detection in VANET. In: International Conference on Advances in Computing and Communications. Berlin, Heidelberg: A Springer-Verlag; 2011

[13] Bißmeyer N et al. Central misbehaviour evaluation for vanets based on mobility data plausibility. In: 9th ACM Workshop on Vehicular Inter-Networking, Systems, and Applications. Low Wood Bay, Lake District, UK; 2012. ACM 978-1-4503-1317

[14] Ismo K et al. Outlier detection using k-nearest neighbour graph. In: Proceedings of the 17th International Conference on Pattern Recognition. 2004

[15] Li W et al. A new intrusion detection system based on KNN classification algorithm in wireless sensor network. Journal of Electrical and Computer Engineering. 2014. pp. 1-8

[16] Maglaras LA. A novel distributed intrusion detection system for vehicular

ad hoc networks. International Journal of Advanced Computer Science and Applications. 2015;**6**:4

[17] Nyakundi E. Using support vector machines in anomaly intrusion detection. PhD Thesis. 2015

[18] Sharafaldin I et al. Towards a reliable intrusion detection benchmark dataset. Software Networking. 2017;**1**:1

[19] Wang L. Support Vector Machines: Theory and Applications. Vol. 177. Berlin Heidelberg: Springer Science & Business Media; 2005

[20] Viriyasitavat W, Boban M, Tsai HM, Vasilakos A. Vehicular communications: Survey and challenges of channel and propagation models. IEEE Vehicular Technology Magazine. 2015;**10**(2):55-66

[21] Bernad'o L, Zemen T, Tufvesson F, Molisch AF, Mecklenbräuker CF. Delay and Doppler spreads of nonstationary vehicular channels for safety-relevant scenarios. IEEE Transactions on Vehicular Technology. 2014;**63**(1):82-93

[22] Guerrero-Ibáñez J, Flores-Cortés C, Zeadally S. Vehicular adhoc networks (VANETs): Architecture, protocols and applications. In: Next-Generation Wireless Technologies. London, U.K.: Springer-Verlag; 2013. pp. 49-70

[23] Zhang R, Cheng X, Yang L, Shen X, Jiao B. A novel centralized TDMA-based scheduling protocol for vehicular networks. IEEE Transactions on Intelligent Transportation Systems. 2015;**16**(1):411-416

[24] Elkan C. Results of the KDD'99 Classifier Learning. ACM SIGKDD Explorations Newsletter. 2000;**1**(2):63

[25] Li Y, Cimini LJ, Sollenberger NR. Robust channel estimation for OFDM systems with rapid dispersive fading channels. IEEE Transactions on Communications. 1998;**46**(7):902-915

[26] Li Y. Pilot-symbol-aided channel estimation for OFDM in wireless systems. IEEE Transactions on Vehicular Technology. 2000;**49**(4):1207-1215

[27] Liang L, Ye H, Li GY. Towards intelligent vehicular networks: A machine learning framework. 2019:1-12. arXiv: 1804.00338v3

[28] Tong W, Hussain A, Bo WX, Maharjan S. "Artificial intelligence for vehicle-to-everything: A survey", special section on emerging technologies on vehicle to everything (V2X). IEEE Access. 2019;**7**:10823-10843

[29] Teh YW. Dirichlet process. In: Encyclopedia of Machine Learning. New York, USA: Springer science+ Business media LLC; 2011. pp. 280-287

[30] Li Y, Cimini LJ. Bounds on the interchannel interference of OFDM in time-varying impairments. IEEE Transactions on Communications. 2001;**49**(3):401-404

[31] Ding W, Yang F, Dai W, Song J. Time-frequency joint sparse channel estimation for MIMO-OFDM systems. IEEE Communications Letters. 2015;**19**(1):58-61

[32] Sutton RS, Barto AG. Introduction to Reinforcement Learning. Vol. 135. Cambridge, MA: MIT Press; 1998

[33] Box GE, Tiao GC. Bayesian Inference in Statistical Analysis. Vol. 40. John Wiley & Sons, University of California; 1992

[34] LeCun Y, Bengio Y, Hinton G. Deep learning. Nature. 2015;**521**(7553): 436-444

[35] Basak D, Pal S, Patranabis DC. Support vector regression. Neural Information Processing. Letters and Reviews. 2007;**11**(10):203-224

[36] Gan G, Ma C, Wu J. Data Clustering: Theory, Algorithms, and Applications. Vol. 20. Philadelphia, PA, United States: Society for Industrial and Applied Mathematics, University City Science Center; 2007

[37] Araniti G, Campolo C, Condoluci M, Iera A, Molinaro A. LTE for vehicular networking: A survey. IEEE Communications Magazine. 2013;**51**(5):148-157

[38] Kenney JB. Dedicated short-range communications (DSRC) standards in the United States. Proceedings of the IEEE. 2011;**99**(7):1162-1182

[39] Abboud K, Zhuang W. Stochastic analysis of a single-hop communication link in vehicular ad hoc networks. IEEE Transactions on Intelligent Transportation Systems. 2014;**15**(5):2297-2307

Chapter 6

Speed Control of Inflow Vehicles for Merging Support on Motorways with Limited I2V Communication

Hisafumi Kokubugata, Hironao Kawashima, Ryotaro Fukui and George Kamata

Abstract

Since merging is one of the most difficult situations for automated driving, merging support provided by infrastructure is extremely helpful for automated vehicles on motorways. However, in the early stages of automated vehicle penetration, infrastructure-side equipment to support merging is limited. In the first half of this chapter, a method to support speed control for inflow vehicles is proposed for the case where there is one detection device on the main lanes and one transmission/detection device for the vehicles in the inflow lane. In the second half, the case equipped with an additional detector and transmitter downstream on each lane is dealt with. New accurate information about the merging vehicles concerned could be given to the speed controller in the infrastructure. The control should be recalculated and sent to the inflow vehicle that has already started. Each control problem is formulated in the optimal control theory and calculated numerically by a nonlinear programming algorithm. The computational experiment is performed in each realistic situation.

Keywords: automated driving, V2I, I2V, merging, speed control, motorway, optimal control, numerical optimization

1. Introduction

Automated driving is entering a new stage. In the United Nations WP29 (World Forum for the Harmonization of Regulations), a proposal was adopted in 2022 to extend automated driving in certain traffic environments (that is Level 3 defined by SAE (Society of Automotive Engineers)) from the limit of 60 km/h to up to 130 km/h (in Ref. [1]). Nevertheless, V2I (Vehicle to Infrastructure)/I2V (Infrastructure to Vehicle)/V2V (Vehicle to Vehicle) communication has not evolved explosively.

Merging is one of the most difficult situations for automated driving on motorways. Previous studies on vehicle speed control in merging areas have mainly focused on the control in situations where V2I/I2V/V2V communication is fully equipped. However, in the early stages of automated vehicle penetration, the infrastructure-side equipment associated with V2I/I2V is limited.

The Strategic Innovation Promotion Program (SIP) of the Japanese national project includes Automated Driving for Universal Services (ADUS, 2014–2023). SIP-ADUS classifies the development of merging support systems into the following four stages:

Day 1: Support for vehicles on the inflow lane (basic).

Detection on main lanes is done at one spot, transmission on inflow lane is made at one spot.

Day 2: Support for vehicles on the inflow lane (advanced).

Detection on main lanes is done in a consecutive section, and transmission on inflow lane is made consecutively.

Day 3: Support for vehicles on both inflow lane and main lanes via I2V.

Day 4: Support for vehicles on both inflow lane and main lanes with full V2I/I2V and V2V.

The situations on Days 1 and 2 are illustrated precisely in Ref. [2]. In the Days 1 and 2 stages, the infrastructure only provides information about vehicles on main lanes to inflow vehicles. It is assumed that vehicle speed in inflow lane is determined by its in-vehicle function.

In this chapter, the situation where the speed control of inflow vehicles is provided by the infrastructure is considered for advanced merging support. In order to determine the control by the infrastructure, a detector has to be added at the same position as the transmitter on inflow lane. This situation is called Day 1^+ by the authors in this chapter. The Day 2^+ is also used in the same manner.

In Section 2, a practical method is proposed that provides speed control to an inflow vehicle for merging under the limited conditions in the Day 1^+ situation.

Next, in Section 3, a modified method that provides speed control to the inflow vehicles under intermediate conditions between Days 1^+ and 2^+, which is called Day 1.5^+ provisionally, is also proposed for near future use. If equipped with an additional detector and transmitter downstream, new accurate information about the merging vehicles concerned could be given to the speed controller in the infrastructure. The control should be recalculated and sent to the inflow vehicle that has already started.

The control problem is formulated in the optimal control theory and calculated numerically by a nonlinear programming algorithm. The computational experiment is performed in each realistic situation associated with Days 1^+ and 1.5^+ respectively.

Finally, in Section 4, the conclusion and perspective are presented. The studies which should succeed in the proposed contents in this chapter are described.

2. Speed control of inflow vehicles for merging support equipped with one detector/transmitter on each lane

The traffic situation in the Day 1^+ stage is illustrated in **Figure 1**. In the following sections, results of computational experiments are provided based on the situations in actual Japanese expressways, and diagrams associated with traffic on the left side are consistently shown in this chapter. In Japan, as in the UK, Australia, etc., traffic keeps to the left side, thus the inflow lane is located on the left side of the main lanes. The lane on the outside of the main lanes is called the outer main lane in this chapter.

The vehicle A passing the transmitter/detector spot in inflow lane receives the speed control computed by the traffic management center or an edge computer on roadside considering the information about the vehicles P, Q, and R already passed through the detector spot in the outer main lane. The position and the speed of each of

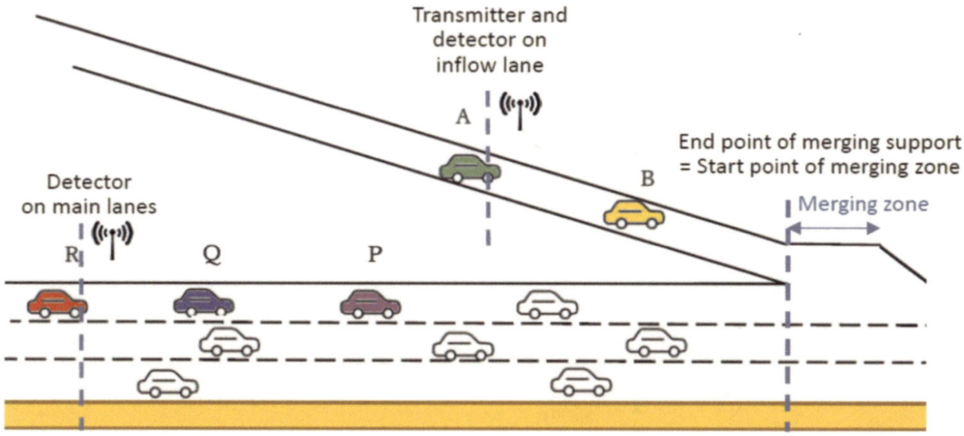

Figure 1.
Schematic illustration of merging support area in Day1⁺ stage.

these vehicles in the merging zone can be estimated from the time and the speed when passing through the detector spot.

Based on all the information, the traffic management center or the edge computer searches for the possible headway between two successive vehicles in the outer main lane where vehicle A can merge into. If the headway between P and Q would be impossible for the merging of A, the next headway between Q and R should be tried. If the appropriate headway is found, the center or the edge computer computes the speed control of vehicle A and sends it to vehicle A. The control support finishes at the start of the merging zone. The merging action itself should be conducted by the in-vehicle automated driving function or the human driver in the Day 1^{+} situation. The method for speed control is described precisely in the next subsection.

2.1 Method of speed control of inflow vehicle

Various methods of control of vehicles have been proposed for merging. Most of them are based on optimization techniques such as dynamic programming [3], game theory [4], and so forth. Recently, deep Reinforcement Learning (deep RL) has been attracting attention in various application fields. A vehicle control method for merging based on the deep RL is also proposed in [5]. Although the deep RL is extremely attractive in the theoretical aspect, each iterative procedure of the RL tries to choose the best action based on consideration of all possible estimated states up to the distant future. Hence, computational efforts expand exponentially and the procedures for implementation are quite complicated.

Computational time should not be ignored in practical applications because the speed control must be applied to the inflow vehicle as soon as possible after receiving the control from the transmitter. Therefore, a simple method of computing speed control is proposed in this section. The method is based on a numerical approach to the traditional optimal control theory.

2.1.1 Scheme of numerical approach to optimal speed control of the inflow vehicle A

A scheme of the numerical approach to optimal speed control of the inflow vehicle A is illustrated in **Figure 2**. The position and the speed of vehicle A at the transmitter/

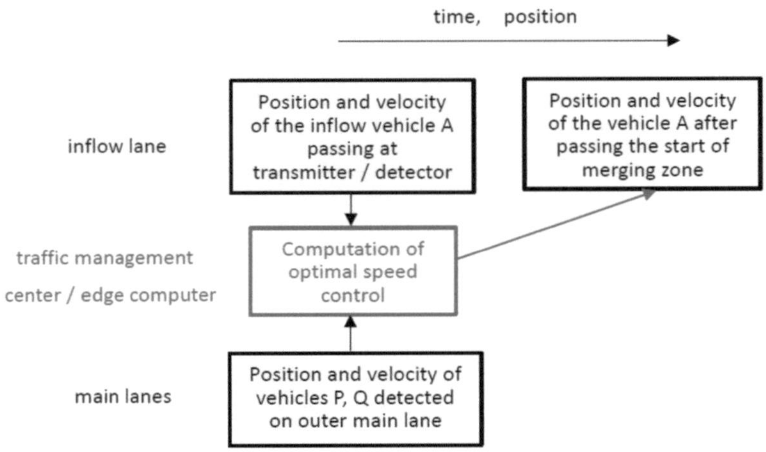

Figure 2.
Scheme of numerical approach to optimal speed control of the inflow vehicle A in Day 1⁺ stage.

detector and the position and the speed of vehicles P and Q detected in the outer main lane are used to calculate the position and the velocity of vehicle A after passing the start of the merging zone. The velocity of vehicle A on the inflow lane is usually lower than the running speed of vehicles on main lanes. It is assumed that the preceding vehicle B in the inflow lane does not prevent vehicle A.

2.1.2 Motion equation of vehicle A

Let
$x(k)$,$v_x(k)$: position (m) and velocity(m/s) of the vehicle A at time *k*: state variables;
$u_x(k)$: acceleration (m/s^2) of vehicle the A at time *k*: control variable;
t_s: sampling time.

$$\boldsymbol{x}(k) = [x(k) \quad v_x(k)]^T$$
$$A = \begin{bmatrix} 1 & t_s \\ 0 & 1 \end{bmatrix}, B = [0 \quad t_s]^T$$

Then, discrete the motion equation of vehicle A is described in (1).

$$\begin{cases} x(k+1) = x(k) + t_s v_x(k) \\ v_x(k+1) = v_x(k) + t_s u_x(k) \end{cases} \tag{1}$$

The motion equation Eq. (1) can be expressed as discrete-time state-space representation of a linear system Eq. (2).

$$\boldsymbol{x}(k+1) = A\boldsymbol{x}(k) + Bu_x(k). \tag{2}$$

2.1.3 Precise formulation of speed control based on optimal control theory and numerical approach

From Eq. (1), when the final time is set as *n*, the entire motion equation of vehicle A is expressed as the following Eq. [3].

$$\begin{cases} x(1) - x(0) - t_s v_x(0) = 0 \\ x(2) - x(1) - t_s v_x(1) = 0 \\ \vdots \\ x(n) - x(n-1) - t_s v_x(n-1) = 0 \\ v_x(1) - v_x(0) - t_s u_x(0) = 0 \\ v_x(2) - v_x(1) - t_s u_x(1) = 0 \\ \vdots \\ v_x(n) - v_x(n-1) - t_s u_x(n-1) = 0 \end{cases} \tag{3}$$

The initial condition of Eq. (3) is set as follows:

$$x(0) = x_0, v(0) = v_0 (k = 0, \cdots, h) \tag{4}$$

where x_0, v_0 are the initial position and velocity of vehicle A at passing the transmitter/detector, and h is the time when the in-vehicle controller in the vehicle A can start operation after receiving the control information.

The limitation of velocity of vehicle A is

$$0 \leq v_x(k) \leq v_{\max} (k = 0, \cdots, n) \tag{5}$$

where $v_{\max}$ is the maximal speed of vehicles A.

Moderate acceleration and deacceleration are imposed as follows (2 m/s$^2 \approx$0.2G):

$$-2 \leq u_x(k) \leq 2 \ (k = 0, \cdots, n) \tag{6}$$

Let

$p(k), q(k)$: the estimated position of vehicle P, Q at time k.

$v_p(k), v_q(k)$: the estimated velocity of vehicle P, Q at time k,

then additional constraints are set as follows:

$$q(k) + f \leq x(k) \leq p(k) - e \ (k = m+1, \cdots, n) \tag{7}$$

where e is safe headway to the preceding vehicle P and f is safe headway to the succeeding vehicle Q, while m is the time when the vehicle A arrives the start of the merging zone.

$$v(k) = v_p(k)(k = m+1, \cdots, n) \tag{8}$$

The velocity of the vehicle A must be equal to the velocity of vehicle P after m.

Subject to the constraints Eqs. (3)–(8), the following evaluation function Eq. (9) is to be minimized.

$$f = -\sum_{k=0}^{n} x(k) + a \sum_{k=0}^{n-1} u_x(k)^2 + b \sum_{k=0}^{n-2} \{u_x(k+1) - u_x(k)\}^2 \tag{9}$$

The first term in Eq. (9) is to move the vehicle A forward as much as possible and the second and the third terms are to make the motion of the vehicle A moderate.

This problem belongs to nonlinear programming problems. Because it is difficult to solve it analytically in general, numerical approaches are usually adopted.

2.2 Computational experiment

In this subsection, a computational experiment in the situation in the Day 1^+ stage is presented.

2.2.1 Environment and conditions in the computational experiment

In 2019, the field operational tests on merging support in the Day 1 stage were conducted by SIP-ADUS at the Airport West on-ramp in Tokyo Metropolitan Expressway. In the test and its preliminary investigation, the allowable positions of the detector and transmitter were specified. Within the permissible limits, the distance between the detector on the main lanes and the start of the merging zone is set to 160 m, while the distance between the transmitter/detector on inflow lane and the start of the merging zone is set to 95 m in this computational experiment. The situation is illustrated in **Figure 3**. In the area, the number of main lanes is two.

Because in the Day 1 situation, vehicles running on main lanes are detected only once at the detector on main lanes, velocity of each vehicle is assumed to be constant throughout the control period. Parameters and initial values in the computational experiment in the Day 1^+ are listed in **Table 1**. The x-axis is set along the road and the positive direction is due to the vehicle running. Its origin point is the start of the merging zone.

In the computational experiments in this chapter, a programming and numeric computing platform MATLAB is used to find the solution to the problem. The discrete optimal control problem Eqs. (3)–(9) belongs to the constrained nonlinear programming problems. Therefore, the optimization tool 'fmincon' in 'Optimization Toolbox' is used. Moreover, because the constraints Eqs. (3)–(8) are linear and the evaluation function Eq. (9) is quadratic, the sequential quadratic programming algorithm 'sqp' is adopted to find the solution.

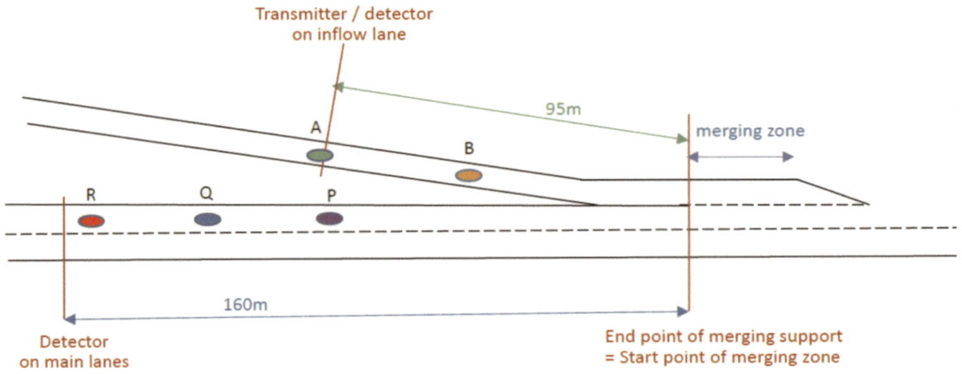

Figure 3.
Positions of detectors and transmitters equipped in Day 1^+ stage.

Item	Attribute value
Time increment	0.1 s
Initial position and velocity of vehicle A	$x_0 = -95$ m,$v_0 = 40$ km/h
Maximal speed of vehicles	$v_{\max} = 60$ km/h
Safe headway to the preceding vehicle P and the succeeding vehicle Q	$e = f = 16.7$ m
Initial positions of vehicle P, Q, and R	$p(0) = -85$ m,$q(0) = -120$ m $r(0) = -155$ m,
Velocity of vehicle P, Q, and R	$v_p(k) = v_q(k) = v_r(k) \equiv 60$ km/h[†]
Delay of transmission[‡]	1.3 s, i.e., $h = 13$
Period of control	12 s, i.e., $n = 120$

[†]*Because vehicles on main lanes are detected only once, the velocity must be dealt with constant.*[‡]*Delay between receiving control information through the in-vehicle antenna and actuation of the in-vehicle control unit.*

Table 1.
Parameters and initial value set in the computational experiment on Day 1[*].

An initial solution has to be given to the iterative optimization procedure. In this computational experiment,

$$v_x(k) = 40\text{km/h}, u_x(k) = 0 \ (k = 0, \cdots, n) \tag{10}$$

these values were adopted for the initial solution. Of course, it is infeasible. The procedure starts at the infeasible initial solution and proceeds to the feasible region and then searches for the optimal solution.

2.2.2 Result of the computational experiment

The result of the computational experiment under the conditions specified above is as follows. In this case, because the initial position of the vehicle P is too far ahead, the merging between P and Q is impossible. Thus, another merging between Q and R is tried and computed, and then it is successfully achieved. The initial positions and the final positions of the vehicles concerned are shown in **Figure 4**.

The trajectory of positions, velocity, and acceleration of vehicles concerned are shown in **Figure 5**. In (b) and (c), graphs associated with vehicle R are omitted because the behavior of vehicle R is assumed to be quite equal to it of vehicle Q.

From (a), the accomplishment of the merging of vehicle A into the headway between vehicles Q and R in the merging zone is observed.

From (b) and (c), it is observed that vehicle A starts at 40 km/h and then deaccelerates for a while, and finally accelerates to 60 km/h before the start of the merging zone. The acceleration in the final phase is to match the speed of vehicles on outer main lane. Due to the acceleration, the initial deceleration is required to adjust the final position of vehicle A.

The speed control of vehicle A that tries merging between the vehicles Q and R are computed in 1.72 s on a note PC (Windows11 with AMD Ryzen 7 5800H 3.2GHz). In the computation shown above, iterations of solution update are limited to 40. When iterations are limited to 10, the computational time decreases to 0.66 s, while the difference in calculation accuracy is only 0.0024%.

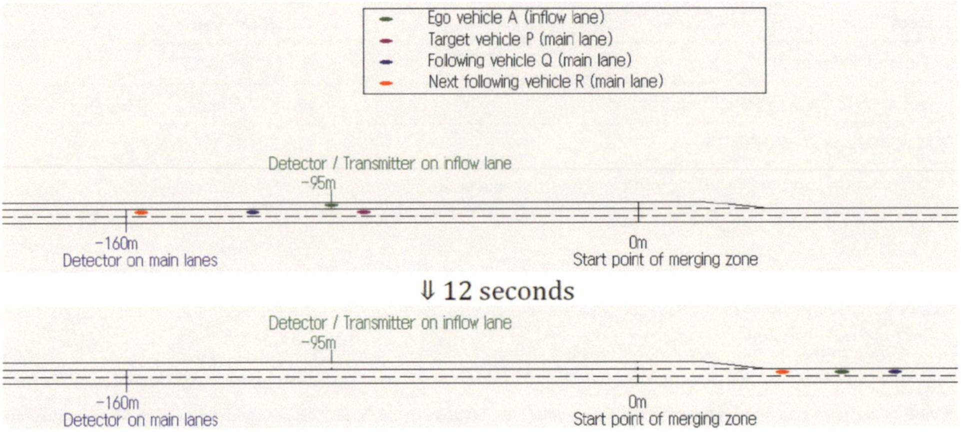

Figure 4.
Initial positions and final positions of vehicles concerned.

2.2.3 Discussions

The result of the computational experiment presented in the previous sub-subsection shows the effectiveness of the method of speed control of inflow vehicles in merging areas in the Day 1^+ stage.

Nevertheless, if the speed of the vehicle on the main line changes significantly due to the slow down around the merging zone after being sensed at the detector, the merging using the computed speed control will not be successful. Generally speaking, the effect of the merging support system in the Day 1^+ stage is limited into the stationary traffic flow.

In order to enhance the effect of merging support, more detectors on main lanes and more detectors/transmitters on inflow lanes are required. The control method of inflow vehicles in the Day 1.5^+ situation is proposed in the following section.

3. Speed control of inflow vehicles equipped with two detectors/transmitters

As explained in Section 1, it is assumed that multiple detectors on main lanes and multiple transmitters/detectors on inflow lanes are equipped in the Day 1.5^+ stage. In **Figure 6**, two detectors on main lanes and two transmitters/detectors on inflow lane are equipped. Based on the additional information provided by the additional facilities, the speed control of the inflow vehicle could be improved.

3.1 Introduction of 'event-driven sequential optimal control'

In the 1980s, the Model Predictive Control (MPC) was proposed and has been applied to various fields. A survey paper including recent applications is presented [6]. Studies on the application of MPC to vehicle control for merging support have been presented, which are based on complete V2I/I2V and V2V communication [7, 8]. The key concept of MPC is that optimal control problems are solved repeatedly at

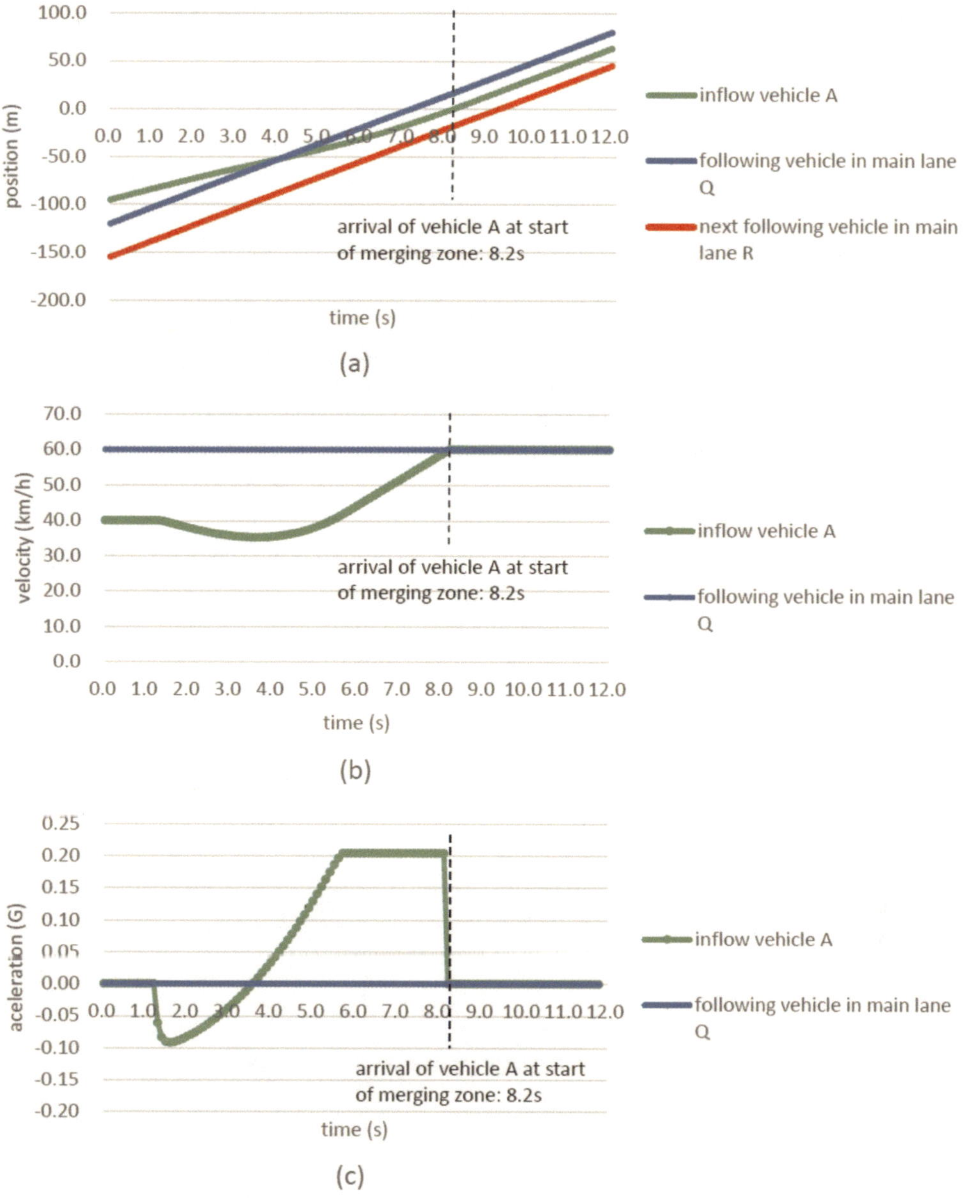

Figure 5.
Positions, velocity, and acceleration of vehicles concerned in the result in Day1$^+$ stage.

certain time intervals under the latest conditions, which are the sum of state transitions due to past control already applied and unexpected changes due to other factors. Although MPC is an effective and powerful control method, computational efforts are considerably large due to repeated optimization.

In the merging situation illustrated in **Figure 6**, an additional control could be computed when new information is detected. Therefore, a two-phase optimization scheme could be applied to the situation. As in this example, the method of recalculating optimal control each time an event occurs is called 'event-driven sequential optimal control' by the authors. The scheme is illustrated in **Figure 7**.

3.2 Computational experiment

In the computational experiment presented in this subsection, most of the environment and conditions are inherited from the experiment shown in Subsection 2.2, in order to focus on the substantial difference between situations/results in Days 1^+ and 1.5^+. The only exception is the exit of the second detectors.

3.2.1 Environment and conditions in the computational experiment

The positions of transmitters and detectors used in the following computational experiment are illustrated in **Figure 8**. Please note that positions of the second transmitter/detector on main lanes and the second detector in inflow lane are determined virtually by the authors, while the first transmitter/detector on main lanes and the first detector in inflow lane are inherited from the previous experiment illustrated in **Figure 3**.

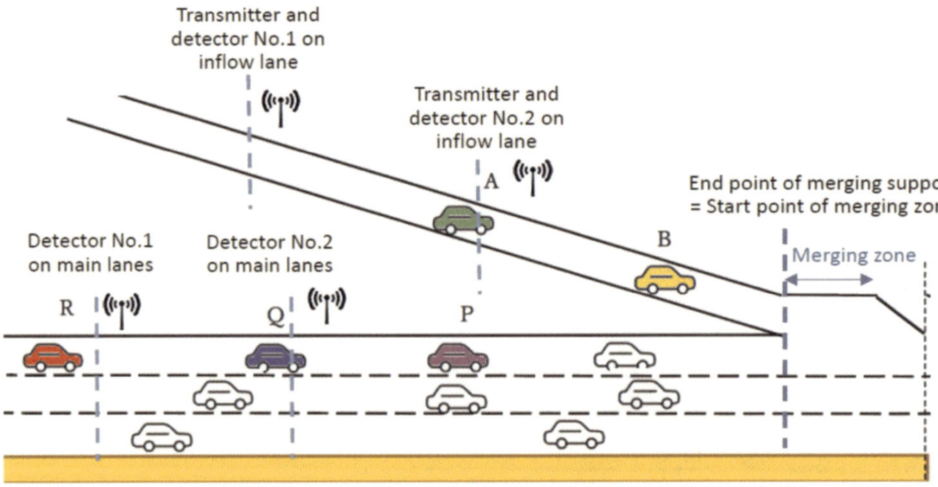

Figure 6.
Merging support area associated with the computational experiment in Day 1.5^+ stage.

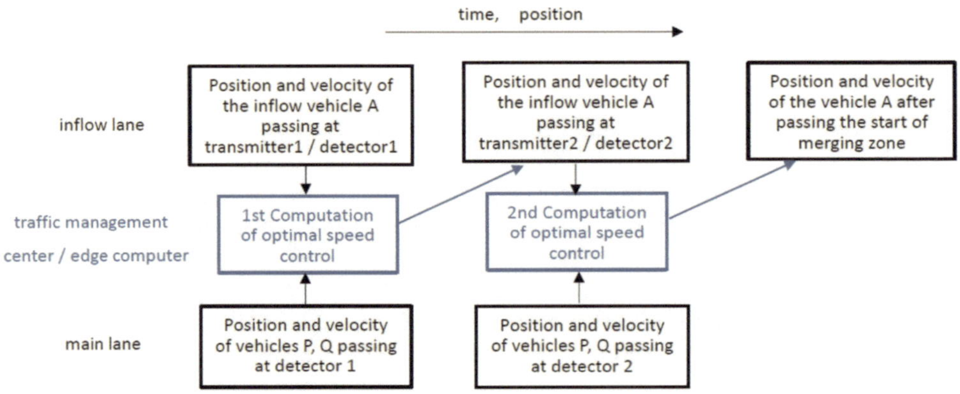

Figure 7.
Scheme of 'event-driven sequential optimal speed control' of the inflow vehicle A in Day 1.5^+ stage.

It is assumed that while running the inflow vehicle A under the speed control computed in Sub-subsection 3.2.2, the current position and velocity of the target vehicle Q passing at the second detector on the outer main lane is informed.

In the following computational experiment, the velocity of the vehicles P, Q, and R passing at the second detector is set at 47.8 km/h. By using the new information, a new speed control for vehicle A is computed and transmitted to it by the second transmitter/detector in inflow lane.

3.2.2 Result of the computational experiment

The result of the computational experiment under the conditions specified above is illustrated in **Figure 9**.

After starting with the initial positions of the vehicles shown in **Figure 4**, new information about vehicle Q is provided at 2.6 s, which is the passing instant of vehicle Q at the second detector. A new computation is executed immediately. In **Figure 9**, data concerned with previous control is expressed as phase 1, while new data concerned with new control computation is expressed as phase 2. (At 2.6 s, vehicle R is not detected yet, and then assumed to follow exactly the move of vehicle Q in phase 2.)

Please note the following points in **Figure 9**:

1. All data in phase 1 are absolutely the same as in **Figure 5**.
2. Trajectories of position, velocity, and acceleration of vehicle Q are discontinuous at 2.6 s due to new detection.
3. The velocity of the vehicle Q has to be constant from 2.6 s because additional detection could not be provided in the Day 1.5 situation.
4. Trajectories of position and velocity of vehicle A are continuous at 2.6 s due to the exact inheritance of data of vehicle A.
5. Velocity of vehicle A are constant for 1.3 s from 2.6 s due to the delay of transmission set in **Table 1**.

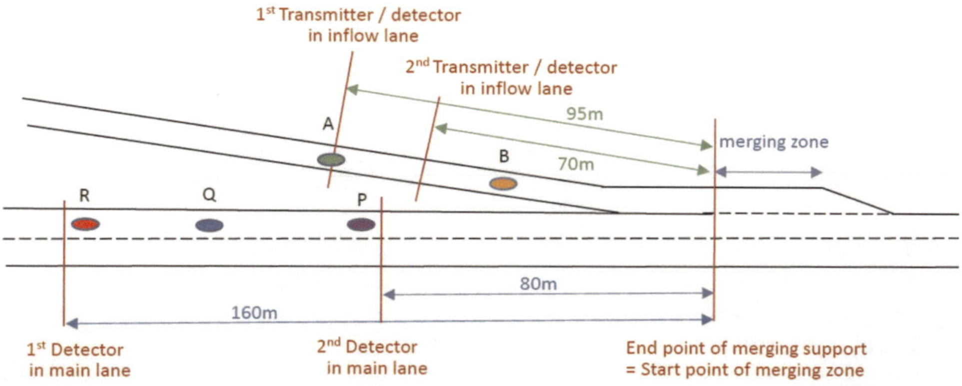

Figure 8.
Positions of detectors and transmitters equipped in Day 1.5⁺ stage.

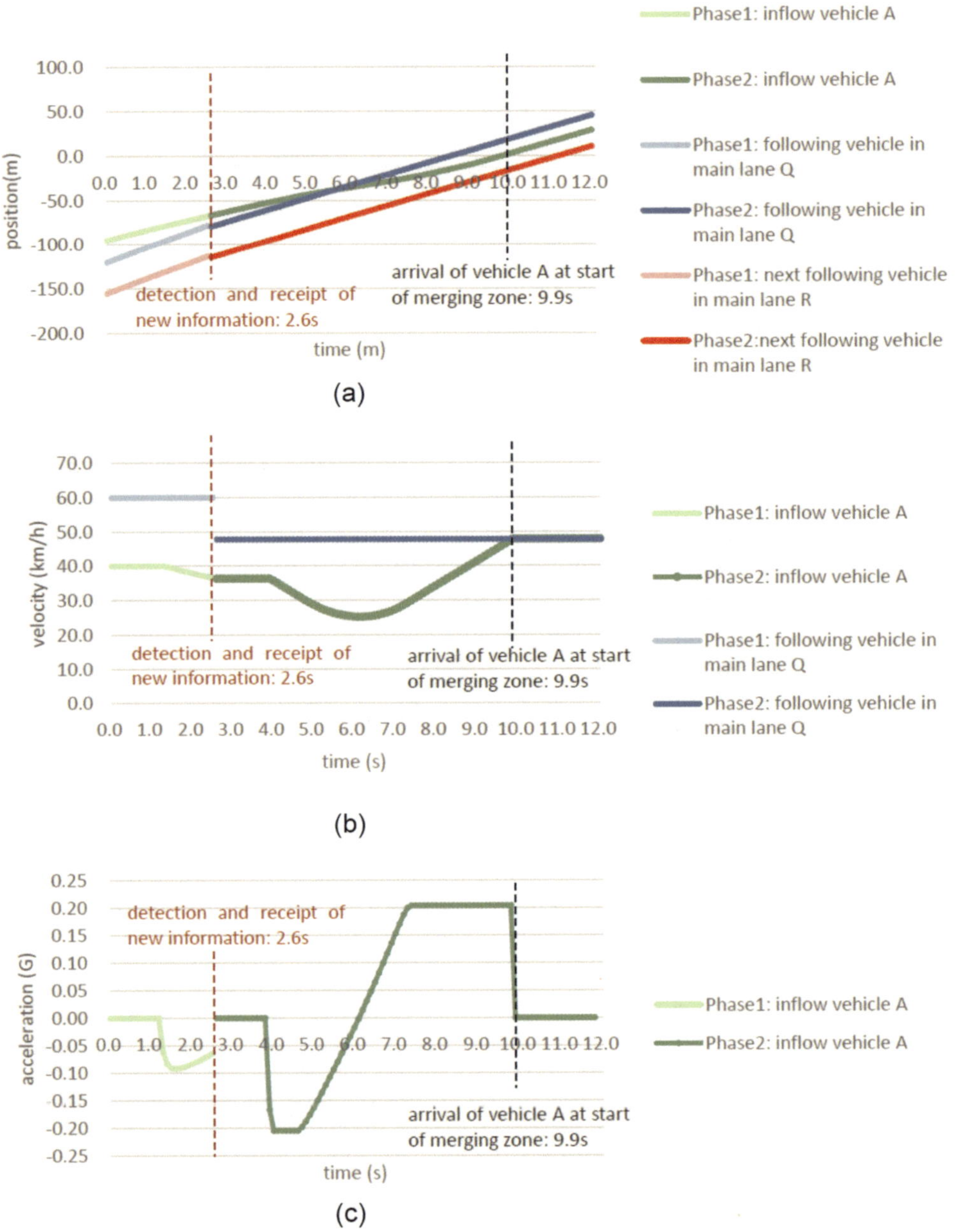

Figure 9.
Positions, velocity, and acceleration of vehicles concerned in the result in the Day 1.5⁺ stage.

6. The timing of acceleration of vehicle A is delayed due to the slower velocity of vehicle Q in the phase 2. As a result, the arrival of the vehicle A at the start of the merging zone is 9.9 s. It delayed 1.6 s compared to the result of the previous result in **Figure 5**.

The computational time is 1.76 s on the note PC (Windows 11 with AMD Ryzen 7 5800H 3.2 GHz). In the computation shown above, iterations of solution update are limited to 40. When iterations are limited to 10, computational time decreases to 0.64 s, while the difference in calculation accuracy is only 0.0033%.

3.2.3 Discussion

In the result of the computational experiment presented in the previous subsubsection, the method of speed control could appropriately cope with the second detection of the vehicle information in outer main lane. Moreover, the computational effort is also suitable for practical use.

The speed control of the inflow vehicle in the Day 1.5$^+$ stage is expected to be improved in comparison with it in the Day1$^+$ stage.

4. Conclusion and perspectives

Provision of the control to inflow vehicles for merging in situations with limited detection and transmission is required during the low penetration of automated driving. In this chapter, speed control methods of the inflow vehicle were proposed for the early stages of penetration in merging support of infrastructure to cope with automated driving development.

In Section 2, the method of speed control of the inflow vehicle was introduced in the situation where one detector for each lane was equipped (Day 1$^+$). The method is formulated as an optimal control formulation and solved by a numerical approach. The computational experiment brings the expected result.

In Section 3, the method of speed control was developed for the application to the situation where an additional detector for each lane was equipped (Day 1.5$^+$). The method belongs to the sequential approach which is composed of two phases of optimization coping with two detectors. The computational experiment also brings the expected result.

Although the performance of speed control with two detectors is of course better than with one, the merging cannot be successful when there are large changes in traffic flow. In order to analyze quantitatively this problem, further computational experiments in various types of traffic flow should be executed. It is expected to conduct verification in a more realistic experiment environment which combines a traffic simulator that reproduces the actual traffic situation with this control method.

In the situation where consecutive detection and transmission are equipped (Day 2$^+$), speed control of inflow vehicles could be solved by Model Predictive Control (MPC). Because MPC needs a large computational effort, some contrivance should be included in the solution procedure.

The main method for merging support is the speed control of inflow vehicles as dealt with in this chapter. In order to proceed to the next stage of merging support (Day 3), collaborative speed control of both inflow vehicles and main lane vehicles must be investigated. To prepare for that, the study on merging support for main lane vehicles is essential, which includes lane change to the inner lane. The control methods proposed in this chapter can be applied also to this problem.

Since communication equipment for automated driving will be expanded by several stages, it is necessary to develop the practical merging support technology in accordance with each stage. The practical methodology should be developed to bridge the gap between theoretical studies assuming that all information is available and the actual communication situation.

Author details

Hisafumi Kokubugata*, Hironao Kawashima, Ryotaro Fukui and George Kamata
Mobility Culture Research Center, Keio University, Kawasaki, Japan

*Address all correspondence to: kokubu@keio.jp

References

[1] UNECE. UN Regulation extends automated driving up to 130 km/h in certain conditions. 2022. Available from: https://unece.org/media/press/368227 [Accessed: 2022 July 27]

[2] Nakagawa T. Merging support service on expressways. 2020. Available from: https://en.sip-adus.go.jp/evt/workshop2020/file/sr/SR_06E_Nakagawa.pdf [Accessed: 2022 July 27]

[3] Sun Z, Huang T, Zhang P. Cooperative decision-making for mixed traffic: A ramp merging example. Transportation Research Part C. 2020; **120**:102764. DOI: 10.1016/j.trc.2020.102764

[4] Fukuyama S. Dynamic game-based approach for optimizing merging vehicle trajectories using time-expanded decision diagram. Transportation Research Part C. 2020;**120**:102766. DOI: 10.1016/j.trc.2020.102766

[5] Liu J, Zhao W, Xu C. An efficient on-ramp merging strategy for connected and automated vehicles in multi-lane traffic. IEEE Transactions on Intelligent Transportation Systems. 2022;**23**(6): 5056-5067. DOI: 10.1109/TITS.2020.3046643

[6] Schwenzer M, Ay M, Bergs T, Abel D. Review on model predictive control: An engineering perspective. The International Journal of Advanced Manufacturing Technology. 2021;**117**: 1327/1349. DOI: 10.1007/s00170-021-07682-3

[7] Cao W, Mukai M, Kawabe T. Two-dimensional merging path generation using model predictive control. Artificial Life Robotics. 2013;**17-3-4**:350/356. DOI: 10.1007/s10015-012-0059-8

[8] Cao W, Mukai M, Kawabe T, Nishira H, Fujiki N. Cooperative vehicle path generation during merging using model predictive control with real-time optimization. Control Engineering Practice. 2015;**34**:98/105. DOI: 10.1016/j.conengprac.2014.10.005